SYMMETRIES AND
SYMMETRY BREAKING
IN FIELD THEORY

SYMMETRIES AND SYMMETRY BREAKING IN FIELD THEORY

P. MITRA

Saha Institute of Nuclear Physics
Calcutta, India

CRC Press
Taylor & Francis Group
Boca Raton London New York

CRC Press is an imprint of the
Taylor & Francis Group, an **informa** business
A CHAPMAN & HALL BOOK

CRC Press
Taylor & Francis Group
6000 Broken Sound Parkway NW, Suite 300
Boca Raton, FL 33487-2742

First issued in paperback 2019

© 2014 by Taylor & Francis Group, LLC
CRC Press is an imprint of Taylor & Francis Group, an Informa business

No claim to original U.S. Government works

ISBN-13: 978-1-4665-8104-3 (hbk)
ISBN-13: 978-0-367-37868-4 (pbk)

Visit the Taylor & Francis Web site at
http://www.taylorandfrancis.com

and the CRC Press Web site at
http://www.crcpress.com

Contents

List of Figures

List of Tables

Preface

While there are many textbooks on quantum field theory today, there seems to be inadequate appreciation of the symmetries involved. Actual courses on field theory have so much material to cover in the formulation and the application of field theory that even topics related to symmetry that are included in good textbooks cannot be sufficiently developed in the classroom or emphasized to students. The present book is an attempt to fill the gap and throw a little more light on symmetry in field theory.

Not all symmetries are realized in full. Some classical symmetries are broken upon quantization. Some symmetries of the theory are broken by the ground state. Such broken symmetries also play very important roles in the modern description of Nature. These topics have naturally been included in the book.

Gravity is definitely described by a field theory and it also has a great deal of symmetry, but it is traditionally kept separate from field theory. This separation has been maintained here. Some readers may look for supersymmetry, but it is too exotic for inclusion. The broad selection of topics is standard, though the presentation may be different in some places.

Students are encouraged to study books on quantum field theory in order to get a bearing on the subject. While this book is based on the published scientific literature, much of it is well known and the original publications would be too scattered to trace. A few references have been included for less familiar areas. I take this opportunity to record my indebtedness to many authors as well as my teachers, collaborators and students.

No attempt has been made to be pedantic. Ideas are presented in a simple way as far as practicable. Those who are familiar with some amount of quantum field theory should be able to follow the mathematical developments in the book, but exposure to elementary particle physics will make the background clearer. A short summary is included in the first chapter.

Saha Institute of Nuclear Physics
Calcutta

Chapter 1

Introduction

1.1   Constituents of matter

Field theory is the setting for modern ideas about the fundamental constituents of matter. These ideas have undergone significant changes since the earliest thinkers visualized atoms as the ultimate building blocks of matter. Chemistry deals with atoms of elements, which mostly occur in combinations called molecules of compounds. But the individual atoms of elements are themselves composite, consisting of electrons moving around nuclei. Electrons are accepted as fundamental at present, but the nuclei are composed of protons and neutrons, which themselves are combinations of entities called quarks. Thus, ordinary matter is supposed to be built out of electrons and quarks.

The forces that we see in nature are as important as the objects which are seen to experience those forces. The most directly perceived force is that of gravity. All matter appears to feel gravitational attraction for all other matter. This is usually described by the concept of a gravitational field throughout space. This field is felt by matter and it is caused by matter. The field is perceived only by its effect on matter, so the phenomenon could instead be described by action at a distance: the action of matter on other pieces of matter. The earth moves in a certain way because of the gravitational attraction of the sun and other bodies and one could directly try to consider those forces. But it is more convenient to visualize a gravitational field at a point, in this case the location of the earth. For a given distribution of all other bodies, or more generally of particles in space, the field indicates the nature of the motion a hypothetical particle at that point would undergo. This is possible because observations have indicated definite laws and precise predictability about the nature of such motion: planets or stars do not move randomly or whimsically on their own. The laws can be formulated in terms of the hypothetical gravitational field. There are laws discovered from observation which tell us how a piece of matter behaves in a gravitational field. There are also laws which tell us how a distribution of matter produces a gravitational field. These together can be said to form the theory of the gravitational field. Newton's ideas led to one such theory. Einstein's ideas have led to a different theory. Both are classical theories in the sense that they do not take quantum theory into account in any way.

There are further examples of field theory corresponding to other forces that are seen in nature. The second kind of force to be discovered can be called electromagnetic. There are forces between charged particles, which may be called electric forces. There are forces on magnets, which may be called magnetic forces, but magnetic forces are caused by electric currents, and these two forces are not essentially different. Just as in the case of the gravitational forces, one can visualize electromagnetic fields, which are a convenient way of representing what kind of force a charged particle would feel at a point. The force involves a velocity independent part, which is essentially the electric field, and a velocity dependent part, which contains the magnetic field. The theory of electromagnetism has to say how a distribution of charges, static or in motion, gives rise to an electromagnetic field. This is in fact given in the form of differential equations by the well known Maxwell equations in classical electromagnetism. A quantum version of this theory is also available, namely, quantum electrodynamics. The classical theory describes light waves and the quantum theory describes quanta, which are quite well known as photons. Photons are not charged and are not acted upon by the electromagnetic field, but carry the forces between charged particles like electrons, quarks, etc.

Additional forces of nature were discovered later. The forces that bind quarks to form protons and neutrons and the latter to form atomic nuclei are much stronger than electromagnetic forces and are referred to as the strong forces. Quarks experience such forces, but not electrons. There is a field describing this strong interaction and it is a complicated analogue of the electromagnetic field which is called the chromodynamic field. The classical version of this theory is of little interest. Quantum chromodynamics involves quanta called gluons, which play a role in gluing the quarks together. While the photons of electrodynamics are uncharged, the gluons of chromodynamics carry chromodynamic charges or color and can be sources of chromodynamic fields in addition to the quarks, which too carry color charges.

There is a fourth kind of interaction between particles. These weak forces are responsible for the radioactive decays of several nuclei, including the neutron itself. The neutron can decay into a proton, emitting an electron and a neutral partner called the (anti)neutrino. This process can be formulated in terms of a transformation of quarks too. The first simple theories of the weak interactions were later replaced by another generalization of electromagnetic field theory. Quanta of the weak field are the W bosons, which are charged. There is also another weak interaction which is more similar to the electromagnetic interaction and has electrically neutral quanta called Z bosons. Both W and Z bosons are heavy, unlike the photon. The weak and electromagnetic interactions have been connected by a coupled field theory of the so-called electroweak interactions.

We can enumerate the particles: quarks, the electron and the neutrino as well as what are called their antiparticles. These are spin 1/2 fermions. There are several kinds of quarks: u, c and t quarks with charge $+\frac{2}{3}$ in units of the electronic charge and d, s and b with charge $-\frac{1}{3}$ in the same units. Similarly, the electron and its partner neutrino too have analogues called the muon with the muon neutrino and the tau with its tau neutrino, all collectively referred to as leptons. These do not take part in the strong interactions. The quark pairs, u and d, c and s, t and b, take part in the electroweak interactions like the lepton pairs. In addition, they participate in the strong interactions. The strong interactions are mediated by gluons and involve color charges. The electroweak interactions are mediated by the photon, the W bosons and the Z boson and involve the electric charge and weak coupling constants. All these mediating particles are bosons with spin 1.

1.2   Symmetries in elementary particle physics

A symmetry is said to be present when two different aspects of an object have the same measure or appearance. We are familiar with reflection symmetry observed in nature and the rotational symmetry exhibited by many geometrical shapes. These symmetries, intuitively understood, can be made mathematically precise. A geometric figure is reflection symmetric about a straight line or a plane if the part to the right, on reflection about the line or the plane, coincides with the part to the left. A geometric figure possesses a rotational symmetry about an axis if a rotation about that axis by a certain angle preserves the totality of the figure. In the case of a circle, the rotation may be through any angle about the axis cutting the plane orthogonally at the center, but in the case of an equilateral triangle or a regular hexagon, for example, the angle of rotation has to be a multiple of a characteristic angle for the figure to be preserved. Thus, symmetries may be continuous or discrete.

In physics, reflections and rotations can certainly be symmetries, but there can be more general symmetries too. One has to specify what is being transformed, how and what remains unchanged under it. For example, if the velocity of a particle is rotated, its kinetic energy does not change. Hamiltonians may be rotationally invariant. The action may be invariant under Lorentz transformations. Maxwell's equations in free space are invariant when $\vec{E}$ is replaced by $\vec{B}$ and $\vec{B}$ by $-\vec{E}$.

In this introductory chapter, we shall mention some celebrated examples of symmetry and symmetry breaking in elementary particle physics.

Like rotational invariance, parity invariance was also believed to hold in nature. Normal electromagnetic interactions certainly do not break parity invariance, which includes the invariance of Hamiltonians under the inversion of all space coordinates and corresponding changes of all fields. However, weak interactions break parity, as suggested by Lee and Yang and demonstrated soon thereafter. The Nobel prize was awarded to them in 1957.

The idea of symmetry in quantum mechanics and in theories of particles and nuclei was developed to a great extent by Wigner, who won a Nobel prize in 1963. Apart from extensive studies of angular momenta, he is known for the theorem that a symmetry operation has to be represented by a unitary operator or an antiunitary one.

Quantum electrodynamics, the quantum version of the gauge invariant electromagnetic theory, took a long time to develop because of technical problems. Feynman, Schwinger and Tomonaga won the Nobel prize for this theory in 1965.

Developments in both theory and experiment followed in the 1960s. Many short lived particles were discovered and classified by their quantum numbers manifested in the strong interactions. Gell-Mann was led to the quark structure of the particles and successfully predicted the existence of some new particle states. He won the Nobel prize in 1969. His discovery was the relevance of the $SU(3)$ group in elementary particle physics. This is simply the group of unitary 3×3 matrices of unit determinant. Among the various quarks that occur, the u and the d are the lightest, and the ones present in the proton and the neutron. If heavier quarks and bound states made out of them are excluded, and the masses of the quarks ignored, an $SU(2)$ symmetry is easy to spot. This is very closely related to the rotation group in a mathematical sense, though of course the transformation from one quark to another is physically very different from a rotation in three dimensional space. This symmetry is called the isospin symmetry in analogy with spin and has been very successful in nuclear physics. The extension of $SU(2)$ to $SU(3)$ becomes natural when the next quark, namely, s, is included too. Its mass is higher, but to some approximation heavier quarks may be neglected and these three species of quarks considered together. The three quarks form the fundamental representation of $SU(3)$, while many mesons, which are quark-antiquark bound states, were seen to be in the adjoint octet representation and some triple quark states in a decuplet representation of the group.

The gauge theory of electromagnetism involves $U(1)$ gauge transformations, where a complex wave function or field may be multiplied by a phase factor, which is a $U(1)$ or 1×1 unitary matrix. It was generalized to theories involving more complicated gauge transformations, like $SU(2)$ or $SU(3)$. This would require fields with internal $SU(2)$ or $SU(3)$ symmetry groups. Weinberg and Salam, following suggestions by Glashow, developed an $SU(2) \times U(1)$ gauge theory to describe electroweak interactions. Doublet representations of $SU(2)$ are provided by the quark pairs u and d, c and s, t and b, as well as by the lepton pairs. Glashow, Salam and Weinberg were awarded the Nobel prize in 1979 after elaborate tests of the model.

The breaking of parity in the weak interactions, which was a surprise in the 1950s, was soon built into theories of weak interactions involving a combination of vector and axial vector currents with equal strength. This makes parity violation maximal. But time reversal invariance was respected by these theories, and there was no experimental evidence to the contrary until the 1960s. Cronin and Fitch were awarded the Nobel prize in 1980 for observing the breaking of this symmetry.

The standard model including both electroweak theory and quantum chromodynamics with its SU(3) color symmetry became established, but some puzzles remained. The major question was how the weak gauge bosons become massive and thereby lead to short range interactions. Apart from the weak gauge boson masses, the masses of other particles too were difficult to explain in the chiral theory of SU(2) involving only left-handed doublets of quarks and leptons. These can be explained by a breaking of the SU(2)×U(1) to U(1) in a special way: through the indirect breaking of the symmetry of the ground state rather than the direct breaking of the symmetry of the interaction. This is called spontaneous symmetry breaking. Spontaneous breaking of symmetry had been studied by Nambu, for which he got a Nobel prize in 2008, shared with Kobayashi and Maskawa, who had studied the mixing of quarks and had noticed that the violation of time reversal needed the existence of at least three pairs or generations of quarks. A detailed theory of spontaneous breaking of gauge symmetry had been worked out in 1964 and was used in the subsequent construction of the standard model of electroweak interactions. It was after the sensational 2012 discovery of a particle which can be regarded as a relic of the spontaneous symmetry breaking of the electroweak gauge symmetry that the theoreticians Englert and Higgs were awarded the Nobel prize in 2013.

We pass on to field theory, symmetries and symmetry breaking in detail.

Chapter 2

Elements of classical field theory

2.1   Functionals

Classical field theory is a generalization of classical mechanics, which deals with discrete particles and a finite number of variables, to situations where one has to consider continuous matter and correspondingly an infinite number of variables. This is actually quite familiar to all who have studied electromagnetism. Although one starts with charged particles, it soon becomes necessary to consider electric fields and magnetic fields which are extended over space. There are equations governing the evolution with time of such fields. The variables of this subject then are the electric and magnetic fields, holding an infinite number of degrees of freedom. Instead of having Lagrangians and Hamiltonians depending on a finite number of generalized coordinates and momenta, one clearly needs a continuum of coordinates and likewise a continuum of momenta. Thus, instead of variables like $q(t)$, one needs variables like $\vec{E}(\vec{x}, t)$. Lagrangians too have to depend on such fields. One thus comes to quantities which depend on functions of space coordinates, in addition to the more common time coordinate.

A functional is a generalization of a function to the case where the argument is not an ordinary real or complex variable but a function. Thus, while x^2 is a function of x and is known when the argument x is given, $\int_0^1 dx\, f(x)$ is a functional of f, which is known when $f(x)$ is known for all x. This is just an example and one can have more complicated cases like $\int_0^\infty dx \exp[-f(x)^2]$ or simpler cases like $f(1) + f(2)$. The latter depends on only two points, $x = 1$ and $x = 2$, but is definitely known if $f(x)$ is known. A functional dependence is indicated by enclosing the function in square brackets: $A[f]$ denotes a functional of f.

A functional changes when the variable function changes, and a derivative may be defined. To specify a derivative, it is necessary to say *where* the function is being changed, so a functional derivative is defined at a point. Thus, one may speak of $\frac{\delta A}{\delta f(y)}$, which indicates how the functional $A[f]$ changes when

f is altered at the point y. This is defined by the limit

$$\frac{\delta A}{\delta f(y)} = \lim_{\epsilon \to 0} \frac{A[f + \epsilon \delta_y] - A[f]}{\epsilon}. \tag{2.1}$$

Here, δ_y is the function defined by the value

$$\delta_y(x) = \delta(x - y). \tag{2.2}$$

Alteration of f at a point y means addition of a function $\epsilon \delta_y$, which vanishes everywhere except at $x = y$. The definition yields the approximate result

$$A[f + \epsilon \delta_y] \simeq A[f] + \epsilon \frac{\delta A}{\delta f(y)} = A[f] + \int dx \frac{\delta A}{\delta f(x)} \epsilon \delta_y(x). \tag{2.3}$$

Addition of further contributions to f leads to the generalization

$$A[f + \delta f] = A[f] + \int dx \frac{\delta A}{\delta f(x)} \delta f(x). \tag{2.4}$$

This relation, or the equivalent relation

$$\delta A[f] = \int dx \frac{\delta A}{\delta f(x)} \delta f(x), \tag{2.5}$$

can also be taken as a definition of the functional derivative.

Just as functions can be of one variable or several variables, functionals can be of functions of several variables or of several functions or of several functions of several variables. In the following section, x will be generalized to the three spatial coordinates or to the four spacetime coordinates. Here f could be something like an electromagnetic field, for example. The functional may be the energy of the field.

2.2   Classical mechanics and classical field theory

A field theory is usually described by a Lagrangian density, which is a local function of fields and their partial derivatives. Just as a Lagrangian L in ordinary mechanics is a function of generalized coordinates and time derivatives, so that one writes $L(q, \dot{q})$, ignoring explicit time dependences, similarly, in a field theory, a Lagrangian is a functional of fields and time derivatives, $L[\phi, \dot{\phi}]$. The functional is usually local, but may include first order spatial derivatives:

$$L = \int d^3 \vec{x} \mathcal{L}(\phi(x), \partial_\mu \phi(x)). \tag{2.6}$$

The local function $\mathcal{L}$ is referred to as the Lagrangian density. In general there may be several fields, but for convenience we display only one field, ϕ. The action is of course

$$S = \int dt\, L = \int d^4x\, \mathcal{L}. \tag{2.7}$$

The equations of motion are obtained by demanding that the action be stationary under variation of the fields with appropriate boundary conditions.

$$\delta S = \int d^4x [(\frac{\partial \mathcal{L}}{\partial \phi})\delta\phi + (\frac{\partial \mathcal{L}}{\partial \partial_\mu \phi})\delta\partial_\mu\phi]. \tag{2.8}$$

Now the variation commutes with partial differentiation, and the partial differentiation in the second term can be shifted by partial integration:

$$\delta S = \int d^4x [(\frac{\partial \mathcal{L}}{\partial \phi})\delta\phi - \partial_\mu(\frac{\partial \mathcal{L}}{\partial \partial_\mu \phi})\delta\phi + \partial_\mu(\frac{\partial \mathcal{L}}{\partial \partial_\mu \phi}\delta\phi)]. \tag{2.9}$$

The last term yields a surface integral, and if $\delta\phi$ is constrained to be zero on this bounding surface, ϕ being fixed at the initial and final time and zero at the boundary of space, the coefficient of $\delta\phi$ in the first two terms yields the equation of motion:

$$\frac{\partial \mathcal{L}}{\partial \phi} = \partial_\mu(\frac{\partial \mathcal{L}}{\partial \partial_\mu \phi}). \tag{2.10}$$

One could also have started from the Lagrangian:

$$\delta S = \int dt\, \delta L(\phi, \dot\phi) = \int d^4x [\frac{\delta L}{\delta \phi}\delta\phi + \frac{\delta L}{\delta \dot\phi}\delta\dot\phi]. \tag{2.11}$$

As δ commutes with time differentiation, the last term can be integrated by parts.

$$\delta S = \int dt\, \delta L = \int d^4x [\frac{\delta L}{\delta \phi}\delta\phi - \frac{\partial}{\partial t}(\frac{\delta L}{\delta \dot\phi})\delta\phi + \frac{\partial}{\partial t}(\frac{\delta L}{\delta \dot\phi}\delta\phi)]. \tag{2.12}$$

The last term can be integrated over time and the integral taken to vanish if $\delta\phi$ vanishes at the initial and final instants, i.e., if ϕ is fixed initially as well as finally. With this condition, the condition for the action to be stationary is that the functional derivative of the action with respect to $\phi(t)$ vanishes, which means that

$$\frac{\delta L}{\delta \phi} = \frac{\partial}{\partial t}(\frac{\delta L}{\delta \dot\phi}). \tag{2.13}$$

This is of the same form as in ordinary mechanics. The only concession to field theory is the appearance of functional derivatives in both terms. To understand this relation better, we examine how L changes when ϕ is altered:

$$\delta L = \int d^3\vec{x} \left[\frac{\partial \mathcal{L}}{\partial \phi} \delta\phi + \frac{\partial \mathcal{L}}{\partial \partial_i \phi} \delta\partial_i\phi + \frac{\partial \mathcal{L}}{\partial \dot{\phi}} \delta\dot{\phi} \right]. \tag{2.14}$$

Using the fact that δ commutes with the space differentiation, we can shift the latter by integration by parts. This produces a surface term at the boundary of space, but this can be taken to be zero if $\delta\phi$ is taken to vanish there, ϕ itself being taken to vanish at the boundary of space. One then obtains the functional derivatives

$$\frac{\delta L}{\delta \phi} = \frac{\partial \mathcal{L}}{\partial \phi} - \partial_i \frac{\partial \mathcal{L}}{\partial \partial_i \phi} \tag{2.15}$$

and the simpler relation

$$\frac{\delta L}{\delta \dot{\phi}} = \frac{\partial \mathcal{L}}{\partial \dot{\phi}}, \tag{2.16}$$

so that the two forms of the equation of motion are identical.

The canonical momentum π is defined by

$$\pi = \frac{\partial \mathcal{L}}{\partial \dot{\phi}} = \frac{\delta L}{\delta \dot{\phi}}. \tag{2.17}$$

A Hamiltonian can be constructed as usual:

$$H = \int d^3\vec{x} \mathcal{H} = \int d^3\vec{x} [\pi\dot{\phi} - \mathcal{L}]. \tag{2.18}$$

As usual, this depends on the momenta and the fields, including spatial derivatives, but not time derivatives, which are removed by the Legendre transformation. Note that

$$\delta H = \int d^3\vec{x} \left[\delta\pi\dot{\phi} + \pi\delta\dot{\phi} - \frac{\partial \mathcal{L}}{\partial \phi}\delta\phi - \frac{\partial \mathcal{L}}{\partial \partial_\mu \phi}\delta\partial_\mu\phi \right]. \tag{2.19}$$

The second term on the right cancels the $\mu = 0$ part of the last term. The remainder of the last term can be integrated by parts and then combined with the third term with the help of the equation of motion to yield

$$\delta H = \int d^3\vec{x} \left[\delta\pi\dot{\phi} - \partial_0 \left(\frac{\partial \mathcal{L}}{\partial \dot{\phi}} \right)\delta\phi \right]. \tag{2.20}$$

This leads to the Hamiltonian equations in nonsingular cases where the π can be varied independently of ϕ and are unconstrained[1] variables:

$$\frac{\delta H}{\delta \pi} = \dot{\phi}, \quad \frac{\delta H}{\delta \phi} = -\dot{\pi}. \tag{2.21}$$

---

[1] If the expressions for the momenta π cannot be solved for the velocities $\dot{\phi}$, there must be relations between the momenta. These relations or constraints occur if $\frac{\partial^2 \mathcal{L}}{\partial \dot{\phi}(\vec{x}) \partial \dot{\phi}(\vec{y})}$ has a vanishing determinant. We shall come across such a situation while studying gauge theories.

The derivatives on the left are functional derivatives in the field theoretic setting. As in the usual particle mechanics situation, one can introduce Poisson brackets of dynamical functionals f, g, and they involve functional derivatives:

$$\{f, g\} = \int d^3\vec{x} \left[\frac{\delta f}{\delta\phi(\vec{x})} \frac{\delta g}{\delta\pi(\vec{x})} - \frac{\delta f}{\delta\pi(\vec{x})} \frac{\delta g}{\delta\phi(\vec{x})} \right]. \tag{2.22}$$

Clearly,

$$\dot{\phi} = \{\phi, H\}, \quad \dot{\pi} = \{\pi, H\}, \tag{2.23}$$

which can be recognized as Hamilton's equations of motion.

Chapter 3

Global symmetries

Global symmetries refer to symmetries under rigid transformations, which may be internal, i.e., resident in internal field space, or associated with space-time transformations like translations, rotations and more generally Lorentz transformations.

3.1 Internal symmetries

Suppose a field theory is characterized by a Lagrangian density $\mathcal{L}(\phi, \partial_\mu \phi)$ and it is invariant under some infinitesimal internal transformations of the fields ϕ

$$\phi \to \phi + \delta\phi = \phi + \epsilon\lambda\phi, \tag{3.1}$$

where ϵ is an infinitesimal parameter and λ is to be understood as a matrix acting on the column vector ϕ. The invariance means that

$$\delta\mathcal{L} = 0. \tag{3.2}$$

But

$$\delta\mathcal{L} = \frac{\partial\mathcal{L}}{\partial\phi}\delta\phi + \frac{\partial\mathcal{L}}{\partial\partial_\mu\phi}\delta\partial_\mu\phi. \tag{3.3}$$

Noting that δ and ∂_μ commute because of the internal nature of the transformation, and using the field equation, we find

$$\partial_\mu\left(\frac{\partial\mathcal{L}}{\partial\partial_\mu\phi}\delta\phi\right) = \delta\mathcal{L}. \tag{3.4}$$

If the Lagrangian density is invariant, the right side vanishes and it follows that

$$\partial_\mu\left(\frac{\partial\mathcal{L}}{\partial\partial_\mu\phi}\lambda\phi\right) = 0. \tag{3.5}$$

This has the form of a conservation law

$$\partial_\mu J^\mu = 0, \tag{3.6}$$

and leads to a conserved charge

$$Q = \int d^3\vec{x}\, J^0 = \int d^3\vec{x}\, \pi\lambda\phi, \tag{3.7}$$

whose time independence follows from the observation that

$$\frac{\partial}{\partial t}J^0 = -\frac{\partial}{\partial x^i}J^i \tag{3.8}$$

and the space integral of the divergence is equivalent to a surface integral at infinity, which can be set equal to zero with the understanding that all fields vanish at infinity.

This connection between a symmetry and a conservation law is generically referred to as Noether's theorem.

Note that the Poisson bracket

$$\{\phi, Q\} = \lambda\phi, \tag{3.9}$$

indicating that the charge is the generator of the symmetry transformation.

We have not specified what λ is. It has to be such that the transformation is a symmetry of the Lagrangian density. For example, if there are two scalar fields ϕ_1, ϕ_2 and the Lagrangian density

$$\mathcal{L} = \frac{1}{2}\partial_\mu\phi_i\partial^\mu\phi_i - \frac{1}{2}m^2\phi_i\phi_i, \tag{3.10}$$

where the sum over i goes over the two values 1 and 2,

$$\delta\phi_1 = \epsilon\phi_2, \quad \delta\phi_2 = -\epsilon\phi_1 \tag{3.11}$$

is a symmetry transformation. λ is a $2 \otimes 2$ antisymmetric matrix in this case.

If there are two symmetries, there will be two charges:

$$Q_1 = \int d^3\vec{x}\, \pi\lambda_1\phi, \quad Q_2 = \int d^3\vec{x}\, \pi\lambda_2\phi. \tag{3.12}$$

In this case, the Poisson bracket

$$\{Q_1, Q_2\} = \int d^3\vec{x}\, \pi[\lambda_1, \lambda_2]\phi, \tag{3.13}$$

as can be seen by using the fundamental Poisson bracket relations among the fields. Thus the commutator of the two λ-s appears in the Poisson bracket. The right side looks like a conserved charge, but it is not obvious at this point that it corresponds to a symmetry transformation. To see that it does, one has to note that the infinitesimal symmetry transformations corresponding to Q_1 and Q_2, namely,

$$\delta\phi = \epsilon_1\lambda_1\phi, \quad \delta\phi = \epsilon_2\lambda_2\phi, \tag{3.14}$$

lead to finite transformations

$$\phi \to \exp(\theta_1\lambda_1)\phi, \quad \phi \to \exp(\theta_2\lambda_2)\phi, \tag{3.15}$$

where θ_1, θ_2 are finite parameters now. The commutator of two such transformations is the composite transformation

$$\phi \to \exp(\theta_1\lambda_1)\exp(\theta_2\lambda_2)\exp(-\theta_1\lambda_1)\exp(-\theta_2\lambda_2)\phi. \tag{3.16}$$

For small θ_1, θ_2, the change is

$$\delta\phi = \theta_1\theta_2[\lambda_1, \lambda_2]\phi. \tag{3.17}$$

This shows the commutator does indeed correspond to a symmetry transformation. Of course, the commutator may vanish, in which case the symmetry is trivial. If it does not, a new symmetry is produced from the two original symmetries and it is generated by the Poisson bracket of the two charges.

3.1.1 Global gauge symmetry

A special internal symmetry transformation that occurs in most theories with fermions is given by

$$\psi \to (1 - i\epsilon)\psi, \tag{3.18}$$

with an infinitesimal parameter ϵ. This is called a global gauge transformation and is of the general form considered in this section. The conserved charge is $\int d^3x \psi^\dagger\psi$, which is just the number operator, and follows from the conserved current

$$\bar{\psi}\gamma^\mu\psi. \tag{3.19}$$

If the fermion field comes in a $U(n)$ multiplet, ϵ can be a hermitian $n \times n$ matrix. Thus one may have a $U(n)$ symmetry. The current (see Chapter 6 for the notation)

$$\bar{\psi}\gamma^\mu T^a\psi \tag{3.20}$$

is conserved in the free theory. If there are gauge fields present, the Lagrangian density is not in general invariant when only the fermion field is transformed.

By evaluating $\delta\mathcal{L}$, one can see that the fermion current satisfies an equation of the form

$$\partial_\mu(\bar{\psi}\gamma^\mu T^a\psi) + g f^{abc} A_\mu^b \bar{\psi}\gamma^\mu T^c\psi = 0. \tag{3.21}$$

Since the covariant derivative appears here, this is referred to as covariant conservation. If there are values of a which are such that f^{abc} vanishes, those components of the currents are indeed conserved. A conserved current also exists because the Lagrangian density remains invariant under a global gauge transformation of the gauge field together with the fermion field. It includes a term containing the gauge field strength which comes from differentiating the Lagrangian density with respect to the derivative of the gauge field:

$$\bar{\psi}\gamma^\mu T^a\psi - f^{abc} A_\nu^b F^{c\mu\nu}. \tag{3.22}$$

3.1.2   Chiral symmetry

Another symmetry occurring in theories with fermions which have *no mass terms* of the usual kind is given by

$$\psi \to (1 - i\epsilon\gamma^5)\psi. \tag{3.23}$$

This is essentially a phase transformation, but the left-handed and right-handed components have different phases; hence it is called a chiral transformation. Mass terms, if present, break this symmetry, which is satisfied by kinetic terms as well as gauge interactions. The corresponding current conservation law is

$$\partial_\mu(\bar{\psi}\gamma^\mu\gamma^5\psi) = 0. \tag{3.24}$$

This equation holds at the classical level only and there are corrections in the quantum theory, as we shall see.

Usually a mass term will be present. In that case, one finds from (3.4) that

$$\partial_\mu(\bar{\psi}\gamma^\mu\gamma^5\psi) = 2im\bar{\psi}\gamma^5\psi. \tag{3.25}$$

This is how the explicit breaking of the symmetry by the mass term manifests itself in a nonconservation equation.

The fermion field here has been assumed to have both left-handed and right-handed components. It is also possible to have a theory with just a left-handed (or a right-handed) fermion without mass. Then the current is of the form

$$\bar{\psi}_L\gamma^\mu\psi_L. \tag{3.26}$$

If the fermion field comes in a $U(n)$ multiplet, ϵ can be a hermitian $n \times n$ matrix. Thus one may have a chiral $U(n)$ symmetry. The current

$$\bar{\psi}\gamma^\mu\gamma^5 T^a\psi \tag{3.27}$$

is conserved in the free theory. If there is a local gauge symmetry, in general these chiral transformations cannot be extended to a symmetry when the fermion has both chiralities. The current is only covariantly conserved:

$$\partial_\mu(\bar{\psi}\gamma^\mu\gamma^5 T^a\psi) + gf^{abc}A_\mu^b\bar{\psi}\gamma^\mu\gamma^5 T^c\psi = 0. \tag{3.28}$$

If only the left (or right) chirality is present, the transformation of the fermion fields together with the corresponding transformation of the gauge fields becomes a gauge symmetry. There is then a conserved current including the gauge fields:

$$\bar{\psi}_L\gamma^\mu T^a\psi_L - f^{abc}A_\nu^b F^{c\mu\nu}. \tag{3.29}$$

The pure fermionic current is again covariantly conserved:

$$\partial_\mu(\bar{\psi}_L\gamma^\mu T^a\psi_L) + gf^{abc}A_\mu^b\bar{\psi}_L\gamma^\mu T^c\psi_L = 0. \tag{3.30}$$

3.2 Translation

Translation invariance of the Lagrangian density means that it is not explicitly dependent on spacetime coordinates. Variation with these coordinates takes place only because of the dependence of the arguments of the Lagrangian density, namely, the fields and their partial derivatives, on the spacetime coordinates.

Therefore,

$$\partial_\mu\mathcal{L} = \frac{\partial\mathcal{L}}{\partial\phi}\partial_\mu\phi + \frac{\partial\mathcal{L}}{\partial\partial_\nu\phi}\partial_\mu\partial_\nu\phi. \tag{3.31}$$

Using the equation of motion in the first term, and noting in the second term that the two partial derivatives on ϕ commute, we get

$$\partial_\mu\mathcal{L} = \partial_\nu(\frac{\partial\mathcal{L}}{\partial\partial_\nu\phi}\partial_\mu\phi). \tag{3.32}$$

The μ derivative on the left can be written in terms of ν:

$$\partial_\nu[\frac{\partial\mathcal{L}}{\partial\partial_\nu\phi}\partial_\mu\phi - \delta_\mu^\nu\mathcal{L}] = 0. \tag{3.33}$$

Setting

$$[\frac{\partial\mathcal{L}}{\partial\partial_\nu\phi}\partial_\mu\phi - \delta_\mu^\nu\mathcal{L}] = \theta^\nu{}_\mu, \tag{3.34}$$

the energy-momentum tensor, we write the above equation as

$$\partial_\nu \theta^\nu{}_\mu = 0, \tag{3.35}$$

a current conservation equation. The conserved charges are

$$P^\mu = \int d^3\vec{x}\,\theta^{0\mu} = \int d^3\vec{x}\,[\pi\partial^\mu\phi - g^{0\mu}\mathcal{L}], \tag{3.36}$$

which are the 4-momenta. Clearly the scalar component P^0 is the Hamiltonian. It follows that

$$\{\phi, P^0\} = \partial^0\phi. \tag{3.37}$$

For $\mu \neq 0$, the above expression for P^μ simplifies to

$$P^i = \int d^3\vec{x}\,\theta^{0i} = \int d^3\vec{x}\,\pi\partial^i\phi, \tag{3.38}$$

so that

$$\{\phi, P^i\} = \partial^i\phi. \tag{3.39}$$

Hence all components satisfy the Poisson bracket relations

$$\{\phi, P^\mu\} = \partial^\mu\phi, \tag{3.40}$$

so that the 4-momenta generate translations in the spacetime coordinates. By integrating the right side of (3.38) by parts, one sees that

$$\{\pi, P^i\} = \partial^i\pi, \tag{3.41}$$

so that it is also true that

$$\{\pi, P^\mu\} = \partial^\mu\pi. \tag{3.42}$$

As P^0 is the Hamiltonian,

$$\{P^\mu, P^0\} = 0. \tag{3.43}$$

Noting (3.38), we find

$$\{P^i, P^j\} = \int d^3\vec{x}\,\pi[\partial^i, \partial^j]\phi = 0, \tag{3.44}$$

because the partial differentiations commute. Thus, in general,

$$\{P^\mu, P^\nu\} = 0. \tag{3.45}$$

Translations commute and their generators commute too.

3.3 Rotations and Lorentz transformations

Rotations — which mix spatial directions — and boosts — which mix space and time directions — are conveniently treated together as general Lorentz transformations. Under such a transformation, the spacetime coordinates transform as

$$x^\mu \to x'^\mu = \Lambda^\mu{}_\nu x^\nu, \tag{3.46}$$

with

$$g_{\mu\rho}\Lambda^\mu{}_\nu\Lambda^\rho{}_\sigma = g_{\nu\sigma}. \tag{3.47}$$

so that proper distances are unchanged.

For infinitesimal transformations

$$\Lambda^\mu{}_\nu = \delta^\mu_\nu + \epsilon^\mu{}_\nu, \tag{3.48}$$

the quadratic relation involving the Λ-s simplifies to

$$\epsilon_{\mu\nu} = -\epsilon_{\nu\mu}. \tag{3.49}$$

Under Lorentz transformations, fields transform as follows:

$$\phi(x) \to \phi'(x'), \tag{3.50}$$

with

$$\phi'(x') = S(\Lambda)\phi(x), \tag{3.51}$$

where $S(\Lambda)$ stands for a representation of the group of Lorentz transformations in the sense that

$$S(\Lambda_1\Lambda_2) = S(\Lambda_1)S(\Lambda_2). \tag{3.52}$$

For infinitesimal transformations, S is of the form

$$S(\Lambda) = 1 + \frac{1}{2}\Sigma^{\mu\nu}\epsilon_{\mu\nu}, \tag{3.53}$$

where Σ is a matrix which can act on the field ϕ considered as a column vector involving an appropriate number of spin components. In the simplest case of a scalar, where ϕ has only a single component, $\Sigma = 0$ and $S = 1$. The matrices Σ have to satisfy commutation relations because of the geometry of rotations and Lorentz transformations. These will be given later in the section.

The change in the field is given by

$$\delta\phi = \phi'(x) - \phi(x) = S(\Lambda)\phi(\Lambda^{-1}x) - \phi(x). \tag{3.54}$$

For infinitesimal Lorentz transformations, this becomes

$$\delta\phi = \frac{1}{2}\Sigma^{\mu\nu}\epsilon_{\mu\nu}\phi - \partial_\mu\phi\epsilon^\mu{}_\nu x^\nu. \qquad (3.55)$$

The change in the Lagrangian density, which is taken to be a scalar, is then

$$\delta\mathcal{L} = -\partial_\mu\mathcal{L}\epsilon^\mu{}_\nu x^\nu = \frac{\partial\mathcal{L}}{\partial\phi}\delta\phi + \frac{\partial\mathcal{L}}{\partial\partial_\rho\phi}\delta\partial_\rho\phi. \qquad (3.56)$$

Using the equation of motion, we can rewrite this as

$$\delta\mathcal{L} = -\partial_\mu\mathcal{L}\epsilon^\mu{}_\nu x^\nu = \partial_\rho(\frac{\partial\mathcal{L}}{\partial\partial_\rho\phi})\delta\phi + \frac{\partial\mathcal{L}}{\partial\partial_\rho\phi}\delta\partial_\rho\phi. \qquad (3.57)$$

If we ignore for the time being the possibility of a noncommutation between δ and ∂_ρ, this can be written as

$$-\partial_\mu\mathcal{L}\epsilon^\mu{}_\nu x^\nu = \partial_\rho(\frac{\partial\mathcal{L}}{\partial\partial_\rho\phi}\delta\phi). \qquad (3.58)$$

This yields

$$-\partial_\mu(\mathcal{L}\epsilon^\mu{}_\nu x^\nu) = -\partial_\rho[\frac{\partial\mathcal{L}}{\partial\partial_\rho\phi}(\partial_\mu\phi\epsilon^\mu{}_\nu x^\nu - \frac{1}{2}\Sigma^{\mu\nu}\epsilon_{\mu\nu}\phi)]. \qquad (3.59)$$

This can be converted to

$$\partial_\rho(g^{\mu\rho}\mathcal{L}\epsilon_{\mu\nu}x^\nu) = \partial_\rho[\frac{\partial\mathcal{L}}{\partial\partial_\rho\phi}(\partial^\mu\phi\epsilon_{\mu\nu}x^\nu - \frac{1}{2}\Sigma^{\mu\nu}\epsilon_{\mu\nu}\phi)] \qquad (3.60)$$

Collecting the ϵ-s, we obtain

$$\partial_\rho[-g^{\mu\rho}\mathcal{L}x^\nu + \frac{\partial\mathcal{L}}{\partial\partial_\rho\phi}(\partial^\mu\phi x^\nu - \frac{1}{2}\Sigma^{\mu\nu}\phi)]\epsilon_{\mu\nu} = 0, \qquad (3.61)$$

or equivalently,

$$\partial_\rho[\theta^{\rho\mu}x^\nu - \frac{1}{2}\frac{\partial\mathcal{L}}{\partial\partial_\rho\phi}\Sigma^{\mu\nu}\phi]\epsilon_{\mu\nu} = 0. \qquad (3.62)$$

Since this holds for arbitrary antisymmetric $\epsilon_{\mu\nu}$, one concludes that

$$\partial_\rho[x^\nu\theta^{\rho\mu} - x^\mu\theta^{\rho\nu} - \frac{\partial\mathcal{L}}{\partial\partial_\rho\phi}\Sigma^{\mu\nu}\phi] = 0. \qquad (3.63)$$

This can be written as

$$\partial_\rho\mathcal{M}^{\rho\mu\nu} = 0, \qquad (3.64)$$

with

$$\mathcal{M}^{\rho\mu\nu} = x^\mu\theta^{\rho\nu} - x^\nu\theta^{\rho\mu} + \frac{\partial\mathcal{L}}{\partial\partial_\rho\phi}\Sigma^{\mu\nu}\phi. \qquad (3.65)$$

This conservation law leads to the conserved charges

$$M^{\mu\nu} = \int d^3\vec{x} \mathcal{M}^{0\mu\nu} = \int d^3\vec{x} [x^\mu \theta^{0\nu} - x^\nu \theta^{0\mu} + \pi\Sigma^{\mu\nu}\phi]. \tag{3.66}$$

These correspond to generalized angular momenta and some of these, namely, those having a time index, are explicitly time dependent, though conserved in time.

3.3.1 Commutation of δ and ∂_μ for Lorentz transformations

As indicated above, δ has been taken to commute with partial differentiation as in the case of translation. In the case of Lorentz transformations, it is not obvious that δ will commute with a partial differentiation because a Lorentz transformation and hence δ act nontrivially on the differentiation operator. So an explicit calculation is useful to check what happens.

Consider the change in a scalar field:

$$\delta\phi = -\epsilon_{\mu\rho}x^\rho\partial^\mu\phi. \tag{3.67}$$

The change in a vector field has an extra term:

$$\delta A_\nu = -\epsilon_{\mu\rho}x^\rho\partial^\mu A_\nu - \epsilon_{\rho\nu}A^\rho. \tag{3.68}$$

From this it is easy to write down the change in the derivative of ϕ:

$$\begin{aligned}
\delta\partial_\nu\phi &= -\epsilon_{\mu\rho}x^\rho\partial^\mu\partial_\nu\phi - \epsilon_{\rho\nu}\partial^\rho\phi \\
&= \partial_\nu\delta\phi.
\end{aligned} \tag{3.69}$$

The change in the derivative of A_ν is

$$\begin{aligned}
\delta\partial_\sigma A_\nu &= -\epsilon_{\mu\rho}x^\rho\partial^\mu\partial_\sigma A_\nu - \epsilon_{\rho\nu}\partial_\sigma A^\rho - \epsilon_{\rho\sigma}\partial^\rho A_\nu \\
&= \partial_\sigma\delta A_\nu.
\end{aligned} \tag{3.70}$$

Thus, in cases of both scalar and vector fields, the operation δ commutes with partial differentiation. In the case of a Dirac field, the change in the field is essentially that of a scalar field together with a spin factor which is in internal spin space and therefore commutes with partial differentiation.

Equivalently, one may consider the Lie derivative of a vector field. For a transformation

$$x^\mu \to x'^\mu = x^\mu + \epsilon\xi^\mu, \tag{3.71}$$

where ξ is a vector field and ϵ a small parameter, one defines

$$\mathcal{L}_\xi A_\mu = \lim_{\epsilon\to 0} \frac{A_\mu(x') - A'_\mu(x')}{\epsilon}. \tag{3.72}$$

Then

$$\mathcal{L}_\xi A_\mu = A_{\mu,\rho}\xi^\rho + A_\rho \xi^\rho_{,\mu}. \tag{3.73}$$

The question is whether

$$\partial_\nu \mathcal{L}_\xi A_\mu \overset{?}{=} \mathcal{L}_\xi \partial_\nu A_\mu \tag{3.74}$$

when ξ corresponds to a Lorentz transformation:

$$\xi^\mu(x) = \epsilon^{\mu\nu} x_\nu. \tag{3.75}$$

Now the left side of (3.74) contains an extra term involving $\xi^\rho_{,\mu\nu}$, which is not present in the right side where only first order derivatives of ξ^ρ occur. This is a possible source of a distinction between the two sides. But the second partial derivative vanishes when ξ^ρ is linear in the spacetime coordinates. Hence the two operations commute for a Lorentz transformation. We have used vector fields, but the same argument applies also to higher order tensor fields because Lie derivatives contain only first order derivatives of ξ^ρ. In the case of a scalar field, the commutation is seen even more simply because the Lie derivative there does not contain any derivative of ξ^μ.

3.3.2   Algebra of the generators

Now we shall study the Poisson brackets of the charges. First of all, we need the fact that

$$\{\phi(\vec{x}), \theta^{0i}(\vec{y})\} = \delta(\vec{x} - \vec{y})\partial^i \phi(\vec{y}). \tag{3.76}$$

To see what happens with θ^{00}, notice that

$$\begin{aligned}
\delta\theta^{00} &= \delta\pi\dot{\phi} + \pi\delta\dot{\phi} - \frac{\partial\mathcal{L}}{\partial\phi}\delta\phi - \frac{\partial\mathcal{L}}{\partial\partial_\mu\phi}\delta\partial_\mu\phi \\
&= \delta\pi\dot{\phi} - \frac{\partial\mathcal{L}}{\partial\phi}\delta\phi - \frac{\partial\mathcal{L}}{\partial\partial_i\phi}\delta\partial_i\phi.
\end{aligned} \tag{3.77}$$

Hence,

$$\{\phi(\vec{x}), \theta^{00}(\vec{y})\} = \frac{\delta\theta^{00}(\vec{y})}{\delta\pi(\vec{x})} = \delta(\vec{x} - \vec{y})\dot{\phi}. \tag{3.78}$$

Thus, for all components,

$$\{\phi(\vec{x}), \theta^{0\mu}(\vec{y})\} = \delta(\vec{x} - \vec{y})\partial^\mu \phi(\vec{y}). \tag{3.79}$$

Hence,

$$\{\phi, M^{\mu\nu}\} = x^\mu \partial^\nu \phi - x^\nu \partial^\mu \phi + \Sigma^{\mu\nu}\phi, \tag{3.80}$$

indicating that

$$\delta\phi = \frac{1}{2}\epsilon_{\mu\nu}\{\phi, M^{\mu\nu}\} \tag{3.81}$$

for an infinitesimal Lorentz transformation. Considering translation and Lorentz generators, one sees that

$$\begin{aligned}\{M^{\mu\nu}, P^i\} &= \int d^3\vec{x}[x^\mu\partial^i\theta^{0\nu} - x^\nu\partial^i\theta^{0\mu} + \Sigma^{\mu\nu}\partial^i(\pi\phi)] \\ &= -g^{i\mu}P^\nu + g^{i\nu}P^\mu\end{aligned} \tag{3.82}$$

by integration by parts. But

$$\{M^{\mu\nu}, P^0\} = \frac{dM^{\mu\nu}}{dt} - \frac{\partial M^{\mu\nu}}{\partial t} = -\frac{\partial M^{\mu\nu}}{\partial t} = -g^{\mu 0}P^\nu + g^{\nu 0}P^\mu. \tag{3.83}$$

Hence, for all components,

$$\{M^{\mu\nu}, P^\rho\} = -g^{\mu\rho}P^\nu + g^{\nu\rho}P^\mu. \tag{3.84}$$

Now we shall try to find expressions for the Poisson brackets of Lorentz generators with one another. Noting that

$$\{\partial^i\phi, M^{\mu\nu}\} = (x^\mu\partial^\nu - x^\nu\partial^\mu)\partial^i\phi + \Sigma^{\mu\nu}\partial^i\phi + (g^{i\mu}\partial^\nu - g^{i\nu}\partial^\mu)\phi \tag{3.85}$$

and

$$\begin{aligned}\{\dot\phi, M^{\mu\nu}\} &= \{\{\phi, P^0\}, M^{\mu\nu}\} \\ &= \{\{\phi, M^{\mu\nu}\}, P^0\} + \{\phi, \{P^0, M^{\mu\nu}\}\} \\ &= (x^\mu\partial^\nu - x^\nu\partial^\mu)\partial^0\phi + \Sigma^{\mu\nu}\partial^0\phi + (g^{0\mu}\partial^\nu - g^{0\nu}\partial^\mu)\phi\end{aligned} \tag{3.86}$$

we can combine these to

$$\begin{aligned}\{\partial^\rho\phi, M^{\mu\nu}\} &= (x^\mu\partial^\nu - x^\nu\partial^\mu)\partial^\rho\phi + \Sigma^{\mu\nu}\partial^\rho\phi + (g^{\rho\mu}\partial^\nu - g^{\rho\nu}\partial^\mu)\phi \\ &= \partial^\rho\{\phi, M^{\mu\nu}\}.\end{aligned} \tag{3.87}$$

Hence,

$$\begin{aligned}\{\phi, \{M^{\mu\nu}, M^{\sigma\tau}\}\} &= \{\{\phi, M^{\mu\nu}\}, M^{\sigma\tau}\} + \{M^{\mu\nu}, \{\phi, M^{\sigma\tau}\}\} \\ &= \{(x^\mu\partial^\nu - x^\nu\partial^\mu + \Sigma^{\mu\nu})\phi, M^{\sigma\tau}\} \\ &\quad + \{M^{\mu\nu}, (x^\sigma\partial^\tau - x^\tau\partial^\sigma + \Sigma^{\sigma\tau})\phi\} \\ &= [(x^\mu\partial^\nu - x^\nu\partial^\mu + \Sigma^{\mu\nu}), (x^\sigma\partial^\tau - x^\tau\partial^\sigma + \Sigma^{\sigma\tau})]\phi \\ &= (g^{\nu\sigma}x^\mu\partial^\tau - g^{\mu\tau}x^\sigma\partial^\nu - g^{\mu\sigma}x^\nu\partial^\tau + g^{\nu\tau}x^\sigma\partial^\mu \\ &\quad - g^{\nu\tau}x^\mu\partial^\sigma + g^{\mu\sigma}x^\tau\partial^\nu \\ &\quad + g^{\mu\tau}x^\nu\partial^\sigma - g^{\nu\sigma}x^\tau\partial^\mu + [\Sigma^{\mu\nu}, \Sigma^{\sigma\tau}])\phi, \tag{3.88}\end{aligned}$$

which is equal to

$$\{\phi, g^{\nu\sigma} M^{\mu\tau} - g^{\mu\tau} M^{\sigma\nu} - g^{\mu\sigma} M^{\nu\tau} + g^{\nu\tau} M^{\sigma\mu}\} \tag{3.89}$$

if the Σ matrices are temporarily assumed to vanish. They do vanish for scalar fields. This suggests that

$$\{M^{\mu\nu}, M^{\sigma\tau}\} = g^{\nu\sigma} M^{\mu\tau} - g^{\mu\tau} M^{\sigma\nu} - g^{\mu\sigma} M^{\nu\tau} + g^{\nu\tau} M^{\sigma\mu}. \tag{3.90}$$

To prove that there are no correction terms involving only ϕ, which has vanishing Poisson brackets with ϕ, one may look for a similar relation involving π. However, the fact that two Lorentz transformations are together equivalent to another Lorentz transformation implies that the Poisson bracket must be expressible in terms of the generators. Hence the above relation must hold without corrections involving fields. Now what happens for nonscalar fields? The same Poisson bracket relation must hold. To ensure this even in the presence of nontrivial Σ matrices, one needs the commutation relations

$$[\Sigma^{\mu\nu}, \Sigma^{\sigma\tau}] = g^{\nu\sigma} \Sigma^{\mu\tau} - g^{\mu\tau} \Sigma^{\sigma\nu} - g^{\mu\sigma} \Sigma^{\nu\tau} + g^{\nu\tau} \Sigma^{\sigma\mu}. \tag{3.91}$$

These commutation relations are indeed necessary for the matrices $S(\Lambda)$ to represent Lorentz transformations. Spin operators have to satisfy the same commutation relations as orbital angular momentum operators. If this equation looks unfamiliar, one can set $\mu = \sigma = 1, \nu = 2, \tau = 3$. Then this takes the form

$$[\Sigma^{12}, \Sigma^{13}] = \Sigma^{23}, \tag{3.92}$$

which resembles the spin commutation relation except for an i. The usual form can be recovered by using $i\Sigma$.

3.4   Symmetric energy-momentum tensor

An action which is translation and Lorentz invariant can be made invariant under general covariant transformations by using general coordinates and introducing additional structure like a metric tensor $g_{\mu\nu}$, which is transformed appropriately under coordinate changes. The measure of integration will now involve a factor $\sqrt{-g}$ so that the correct volume integration is carried out. Apart from that, all derivatives of the fields ϕ have to be expressed in terms of covariant derivatives of the fields. A scalar Lagrangian density can be obtained by contracting such derivatives and other fields wherever necessary by using the inverse $g^{\mu\nu}$ of the covariant metric tensor if there is no spinor field. Thus we write

$$\delta S[g, \phi] = 0 \tag{3.93}$$

under general, infinitesimal coordinate transformations

$$x^\mu \to x'^\mu = x^\mu + \epsilon\xi^\mu. \tag{3.94}$$

The action changes due to the change of the field ϕ and also due to the change of the metric g under this transformation. We write

$$\delta S = \int d^4x \frac{\delta S}{\delta\phi}\delta\phi + \int d^4x \frac{\delta S}{\delta g_{\mu\nu}}\delta g_{\mu\nu}. \tag{3.95}$$

The first term vanishes when the equation of motion of the field is satisfied. The second term is written as

$$-\frac{1}{2}\int d^4x\sqrt{-g}T^{\mu\nu}\delta g_{\mu\nu} \tag{3.96}$$

where $T^{\mu\nu}$ is called the symmetric energy momentum tensor of the field ϕ. Its symmetry follows from the symmetry of $g_{\mu\nu}$, which is preserved under the change induced by the coordinate transformation. The second term can also be written in terms of the Lagrangian density, which enters the action through

$$S = \int d^4x\sqrt{-g}\mathcal{L}(\phi, \phi_{;\mu}, g^{\mu\nu}). \tag{3.97}$$

Considering that

$$\delta\sqrt{-g} = \frac{1}{2}\sqrt{-g}g^{\mu\nu}\delta g_{\mu\nu} \tag{3.98}$$

and

$$\delta g^{\mu\nu} = -g^{\mu\rho}g^{\sigma\nu}\delta g_{\rho\sigma}, \tag{3.99}$$

we can write

$$T^{\mu\nu} = 2g^{\mu\rho}g^{\sigma\nu}\frac{\partial\mathcal{L}}{\partial g^{\rho\sigma}} - g^{\mu\nu}\mathcal{L}, \tag{3.100}$$

which shows that it is a generalization of the canonical energy momentum tensor defined earlier. The change of $g_{\mu\nu}$ is proportional to its Lie derivative

$$\mathcal{L}_\xi g_{\mu\nu} = \xi^\sigma_{;\nu}g_{\mu\sigma} + \xi^\sigma_{;\mu}g_{\nu\sigma} + \xi^\sigma g_{\mu\nu;\sigma} = \xi_{\mu;\nu} + \xi_{\nu;\mu}. \tag{3.101}$$

As $T^{\mu\nu}$ is symmetric,

$$\begin{aligned}
0 &= \int d^4x\sqrt{-g}T^{\mu\nu}(\xi_{\mu;\nu} + \xi_{\nu;\mu}) \\
&= 2\int d^4x\sqrt{-g}T^{\mu\nu}\xi_{\mu;\nu} \\
&= 2\int d^4x\sqrt{-g}[(T^{\mu\nu}\xi_\mu)_{;\nu} - T^{\mu\nu}_{;\nu}\xi_\mu] \\
&= 2\int d^4x[(\sqrt{-g}T^{\mu\nu}\xi_\mu)_{,\nu} - \sqrt{-g}T^{\mu\nu}_{;\nu}\xi_\mu].
\end{aligned} \tag{3.102}$$

Now, if we choose ξ_μ to vanish at infinity, the first integral vanishes, while the arbitrariness of ξ_μ yields the covariant conservation law

$$T^{\mu\nu}_{\ ;\nu} = 0, \qquad (3.103)$$

which reduces to the ordinary conservation law

$$T^{\mu\nu}_{\ ,\nu} = 0 \qquad (3.104)$$

when inertial coordinates are used.

The above analysis is valid when there are no spinor fields. Such fields require tetrads in the action. Tetrads satisfy the relation

$$e^a_\mu e^a_\nu = g_{\mu\nu}, \qquad (3.105)$$

where a is an internal index taking four values. This relation implies that for symmetric $T^{\mu\nu}$,

$$\delta S = -\frac{1}{2} \int d^4x \sqrt{-g} T^{\mu\nu} \delta g_{\mu\nu} = - \int d^4x \sqrt{-g} T^{\mu\nu} e^a_\mu \delta e^a_\nu. \qquad (3.106)$$

The second expression is used to define $T^{\mu\nu}$ for spinor fields, where the action cannot be expressed without introducing tetrads. The action is arranged to be not only generally covariant but also invariant under local Lorentz transformations of the tetrads by using the so-called spin connection generated by the tetrad to construct an appropriate covariant derivative of the spinor field. This invariance,

$$\int d^4x \sqrt{-g} T^{\mu\nu} e^a_\mu e^b_\nu \epsilon_{ab} = 0, \qquad (3.107)$$

which has to hold for arbitrary antisymmetric $\epsilon_{ab}(x)$, ensures that $T^{\mu\nu}$ is symmetric. As a result, the first expression for the energy momentum tensor holds even for spinor fields.

Chapter 4

Local symmetry and constraint theory

In contrast to the symmetries discussed above, where the transformations involve parameters which are independent of spacetime coordinates, some systems possess local gauge symmetry. Here the transformations involve local parameters, namely, ones which can vary with spacetime location. The most common system of this type is the electromagnetic field.

4.1   Electromagnetic field

The electromagnetic theory can be described by Maxwell equations:

$$\nabla.\vec{E} \;=\; 0, \tag{4.1}$$

$$\nabla.\vec{B} \;=\; 0, \tag{4.2}$$

$$\nabla \times \vec{B} \;=\; \dot{\vec{E}}, \tag{4.3}$$

$$\nabla \times \vec{E} \;=\; -\dot{\vec{B}}. \tag{4.4}$$

Only the free theory is described by these equations, which do not contain source charges or currents. These equations are expressed in terms of physical fields. The second and fourth equations suggest the introduction of potentials:

$$\vec{B} \;=\; \nabla \times \vec{A}, \tag{4.5}$$

$$\vec{E} \;=\; -\nabla\phi - \dot{\vec{A}}. \tag{4.6}$$

The fields can be expressed in terms of these potentials, but the latter are not uniquely determined by the physical fields, as is clear from the fact that the transformations

$$\phi \to \phi - \dot{\chi}, \quad \vec{A} \to \vec{A} + \nabla\chi, \tag{4.7}$$

where χ is an arbitrary field, leave $\vec{E}, \vec{B}$ invariant but change the potentials. These are called local gauge transformations and differ from the transformations considered in the preceding chapter through the dependence on a local field instead of a finite number of global parameters.

If the Maxwell equations are to be obtained from a Lagrangian density, it has to be noted that they are first order equations and are of a Hamiltonian rather than a Lagrangian form. Second order equations can be envisaged in terms of the potentials. A Lagrangian density is

$$\mathcal{L} = -\frac{1}{4}(\partial_\mu A_\nu - \partial_\nu A_\mu)(\partial^\mu A^\nu - \partial^\nu A^\mu), \tag{4.8}$$

where A_μ is the 4-vector built out of $\phi, \vec{A}$. The equation of motion following from this equation is

$$\partial_\mu(\partial^\mu A^\nu - \partial^\nu A^\mu) = 0, \tag{4.9}$$

which can be identified with the first and the third Maxwell equations using the correspondence

$$\partial^\mu A^\nu - \partial^\nu A^\mu = F^{\mu\nu}, \tag{4.10}$$

$$F^{i0} = E^i, \quad F^{ij} = -\epsilon_{ijk}B^k. \tag{4.11}$$

This Lagrangian density is invariant under the above local gauge transformation, which can also be written as

$$A_\mu \to A_\mu + \partial_\mu \chi. \tag{4.12}$$

One can try to go to a Hamiltonian formulation from this Lagrangian density, but the first observation that one makes is that

$$\pi_0 \equiv \frac{\partial \mathcal{L}}{\partial \dot{A}^0} = 0, \tag{4.13}$$

because the time derivative of A_0 does not occur in the Lagrangian. The other momenta are well defined:

$$\pi_i \equiv \frac{\partial \mathcal{L}}{\partial \dot{A}^i} = -E^i. \tag{4.14}$$

The vanishing of a momentum (field) means that canonical Poisson brackets cannot be used here. The vanishing has to be treated as a constraint in phase space. A theory of constraints is available[1] to handle such cases of singular Lagrangians where det $\frac{\partial^2 L}{\partial \dot{q}_i \partial \dot{q}_j} = 0$, so that the velocities cannot be solved for in terms of momenta.

---

[1] H. J. Rothe and K. D. Rothe, *Classical and Quantum Dynamics of Constrained Hamiltonian Systems*, World Scientific, Singapore (2010).

4.2 A constrained system in mechanics

Consider a simple mechanical system described by the Lagrangian

$$L = \frac{1}{2}(\dot{x} - z)^2 + \frac{1}{2}(\dot{y} - z)^2 - \frac{1}{2}(x - y)^2 \tag{4.15}$$

involving three generalized coordinates x, y, z. If we try to construct the Hamiltonian, we first notice that $\dot{z}$ does not occur in the Lagrangian, resulting in a constraint

$$p_z = 0. \tag{4.16}$$

This equation signifies that one cannot solve for $\dot{z}$ in terms of p_z and arises because $\frac{\partial^2 L}{\partial \dot{z}^2} = 0$ or more generally because the determinant $\frac{\partial^2 L}{\partial \dot{q}_i \partial \dot{q}_j} = 0$. Such constraints imply restrictions in phase space. We shall see gradually how one can take them into account.

Note that the other momenta are well defined:

$$p_x = \dot{x} - z, \quad p_y = \dot{y} - z. \tag{4.17}$$

The Hamiltonian can be written as

$$H = p_x \dot{x} + p_y \dot{y} - L = \frac{1}{2}(p_x^2 + p_y^2) + z(p_x + p_y) + \frac{1}{2}(x - y)^2, \tag{4.18}$$

where the term $p_z \dot{z}$ has been dropped because $p_z = 0$. The Hamiltonian dictates the evolution of the system in time. As there is a constraint, it will evolve in time, and for consistency, one requires the time derivative of the constraint to vanish. This yields

$$0 = \{p_z, H\} = -p_x - p_y. \tag{4.19}$$

This is a constraint too, and it is new, i.e., independent of the previous constraint. One can check that

$$\{p_x + p_y, H\} = 0. \tag{4.20}$$

Thus there is no further constraint required for consistency in time. So this system with three generalized coordinates has two constraints. The constraint $p_z = 0$ is called a primary constraint because it appeared at the first stage, while the constraint $p_x + p_y = 0$ is called a secondary constraint because it appeared at a later stage following the demand of consistency in time. An important fact is that

$$\{p_x + p_y, p_z\} = 0 : \tag{4.21}$$

the constraint functions have vanishing Poisson brackets with one another.

When this happens in a system, one says that the constraints are first class. If Poisson brackets of constraint functions do not vanish, they are said to be second class; more precisely, if there is a set of constraint functions whose Poisson brackets form a matrix with a nonzero determinant, they are second class. If some subdeterminants are nonzero but the determinant vanishes, there may be a mixture of first and second class constraints.

First class constraints arise in systems with local gauge invariance. To see that the mechanical system possesses such an invariance, note that the Lagrangian does not change under the transformations

$$x \to x + a(t), \quad y \to y + a(t), \quad z \to z + \dot{a}. \qquad (4.22)$$

This is a gauge transformation in the sense that a is an arbitrary function of time. It has the consequence that the time dependence of the coordinates cannot be determined by the equations of motion. This is very similar to what happens in electrodynamics with the potentials as variables. It means that the generalized coordinates x, y, z are not the proper physical degrees of freedom of the system. To identify physical variables, one has to get rid of the constraints. In the present case, both constraints involve momenta. Since phase space variables come in pairs, these momenta have to be removed along with appropriate generalized coordinates. When two pairs of canonical variables are eliminated from the original three, only one pair can remain as physical.

Note in this connection that $x - y$ is invariant under the gauge transformation. Furthermore, both $x - y$ and p_x (say) have zero Poisson brackets with the two constraints. They may be regarded as the true coordinates of the physical phase space. The Hamiltonian may be written in view of the constraints as

$$H = p_x^2 + \frac{1}{2}(x - y)^2. \qquad (4.23)$$

However, the choice of these coordinates is not unique.

4.3   Electromagnetic field without gauge fixing

This kind of analysis can be attempted in the case of electrodynamics. Let us first construct the Hamiltonian.

$$H = \int d^3x [\frac{1}{2}(E^2 + B^2) - A_0 \nabla.\vec{E}]. \qquad (4.24)$$

Using this, one has to examine the consistency of the primary constraint $\pi_0 = 0$ with time evolution.

$$0 = \{\pi_0, H\} = \nabla.\vec{E}. \qquad (4.25)$$

Thus here too there is a secondary constraint, which we identify immediately as Gauss' law. $\nabla.\vec{E}$ has a zero Poisson bracket with H, so there is no further constraint. The Poisson bracket

$$\{\pi_0, \nabla.\vec{E}\} = 0, \tag{4.26}$$

so that we have first class constraints. This is in accord with the gauge invariance of the electrodynamic theory.

To eliminate the constrained variables from the phase space, one needs conjugates of the constraints. For π_0, the canonical conjugate is A^0. However, for the Gauss' law constraint, there is no obvious choice. Even without making a choice, it is possible to see that four phase space variables are to be eliminated from the eight original phase space variables – the fields A^μ and their momenta. This leaves four phase space variables, corresponding to only two fields. As A^0 is already among the eliminated, it is clear that out of the three spatial components of the potential, only two are physical, but which two? The physical variables have to have vanishing Poisson brackets with the constraints and thus be gauge invariant. This is true for the transverse components of the vector. One can write

$$\vec{A} = \vec{A}_T + \vec{A}_L, \tag{4.27}$$

where the two parts, transverse and longitudinal, satisfy

$$\nabla.\vec{A}_T = 0, \quad \nabla \times \vec{A}_L = 0. \tag{4.28}$$

By applying the differential operator $\nabla \times (\nabla \times ..)$ on $\vec{A}$, one finds

$$\vec{A}_T = -\frac{\nabla(\nabla\cdot) - \nabla^2}{\nabla^2}\vec{A}. \tag{4.29}$$

Clearly, this separation is nonlocal. A gauge transformation changes $\vec{A}$ by a gradient, which alters its longitudinal part only. The transverse part is gauge invariant. Note that the magnetic field is

$$\vec{B} = \nabla \times \vec{A}_T \tag{4.30}$$

as the longitudinal part drops out. For a free electromagnetic field,

$$\vec{E} = \vec{E}_T = -\dot{\vec{A}}_T, \tag{4.31}$$

while the longitudinal part, which is related to the scalar potential and the longitudinal part of the vector potential, vanishes. Thus, in this case, the Hamiltonian is expressed purely in terms of physical fields. The Poisson brackets are not canonical because one has to take the nonlocal, transverse part of $\vec{A}$.

$$\{A_T^i(\vec{x}), E^j(\vec{y})\} = (-\delta^{ij} + \frac{\partial^i \partial^j}{\nabla^2})\delta(\vec{x} - \vec{y}) \equiv -\delta_T^{ij}(\vec{x} - \vec{y}). \tag{4.32}$$

In the presence of sources, the longitudinal electric field is determined by those sources.

4.4   A system with second class constraints

The mechanics model considered above has first class constraints and local gauge invariance. A modified version of the system can be obtained by adding a term which breaks the gauge invariance, say $-z^2$. Then the momenta are unchanged, but the Hamiltonian becomes

$$H = \frac{1}{2}(p_x^2 + p_y^2) + z(p_x + p_y) + \frac{1}{2}(x - y)^2 + z^2. \tag{4.33}$$

The secondary constraint changes to

$$0 = \{p_z, H\} = -p_x - p_y - 2z. \tag{4.34}$$

This has a nonzero Poisson bracket with the primary constraint p_z, so that one has second class constraints here. Correspondingly, there is no gauge invariance present any more.

Only one pair of canonical variables is unphysical here: they may be taken to be z, p_z and the Hamiltonian rewritten by expressing z using the secondary constraint:

$$H = \frac{1}{4}(p_x - p_y)^2 + \frac{1}{2}(x - y)^2. \tag{4.35}$$

These may be supplemented by canonical Poisson brackets for the physical variables, which are x, y, p_x, p_y. This may be surprising because x, for example, does not have a vanishing Poisson bracket with the secondary constraint. But in the presence of second class constraints, one has to introduce new brackets, called Dirac brackets, which are consistent with the constraints. For any dynamical variables, one sets

$$\{A, B\}_D = \{A, B\} - \sum_{ij}\{A, C_i\}\{C, C\}_{ij}^{-1}\{C_j, B\}, \tag{4.36}$$

where D denotes the Dirac bracket and the sum is over the full set of constraint functions C_i. The inverse is to be taken in the matrix sense, i.e., the Poisson bracket matrix of the set of constraint functions is to be computed and the inverse constructed.

With this definition, it can be seen that constraint functions have vanishing Poisson brackets with all variables. In the case at hand, one can check that the physical variables have Dirac brackets identical with Poisson brackets.

4.5   Electromagnetic field with gauge fixing

The two constraints arising in electromagnetic theory are first class constraints because of the gauge invariance. One can supplement them by additional constraints to make the full set second class. This is done so as to help us decide which variables to eliminate as unphysical. As indicated above, there is some arbitrariness in this choice. For instance, one can choose a constraint

$$A_0 = 0 \tag{4.37}$$

to remove A_0 along with π_0. But this is not enough. One needs another constraint to go with Gauss' law. For instance, one may choose

$$\nabla.\vec{A} = 0 \tag{4.38}$$

as the second constraint introduced by hand. These additional constraints are called gauge conditions and the procedure is called gauge fixing. Sometimes only one gauge condition is introduced and a second one generated by some procedure from the first. In the Hamiltonian formalism one needs these extra constraints to make the full set second class. The choice of the gauge conditions is arbitrary, but they have to make the set of constraints, together with gauge conditions, second class.

With the present choice of gauge conditions, the matrix of Poisson brackets is block diagonal, because the first gauge condition goes with the primary constraint and the second one with the secondary constraint. So the inverse matrix is also block diagonal. Hence we have, for instance,

$$
\begin{aligned}
\{A^i(\vec{x}), E^j(\vec{y})\}_D &= -\delta^{ij}\delta(\vec{x}-\vec{y}) - \int d^3\vec{w} \int d^3\vec{z} \{A^i(\vec{x}), \nabla.\vec{E}(\vec{z})\} \\
&\quad \{C,C\}^{-1}_{\nabla.\vec{E}(\vec{z})\ \nabla.\vec{A}(\vec{w})} \{\nabla.\vec{A}(\vec{w}), E^j(\vec{y})\} \\
&= -\delta^{ij}_T(\vec{x}-\vec{y}).
\end{aligned}
\tag{4.39}
$$

In this formulation, one can keep all three components of the vector potential, but the Dirac bracket picks out only the transverse part. Other gauge conditions can of course be chosen, and Dirac brackets will not always be of this simple form. For instance, Gauss' law may be given a partner $A_3 = 0$. This is not rotationally invariant, but has certain advantages, notably that it has no derivatives.

An alternative approach to quantizing the electromagnetic theory is to break the gauge invariance by adding a gauge noninvariant term in the Lagrangian, in the way the model of the previous section was obtained by the addition of $-z^2$. This is often done in the functional integral formalism.

Chapter 5

Functional integral formulation of field theory

In the standard formulation of quantum mechanics, observables are represented by operators in a Hilbert space and states by vectors in it. Quantum field theory too is formulated in terms of field operators. States in field theory include the vacuum state, one particle states, two particle states and so on. One does not measure operators or state vectors, however; measurements involve eigenvalues of operators or transition probabilities between states. It turns out that one can formulate quantum mechanics and field theory without using operators. Indeed, the probability amplitude for a transition from a position q_1 at time t_1 to a position q_2 at time t_2 can be expressed as a *path integral*

$$\int \mathcal{D}q(t) e^{\frac{i}{\hbar} \int_{t_1}^{t_2} dt \ L(q(t),\dot{q}(t))}, \tag{5.1}$$

where $q(t)$ runs over all possible functions satisfying the conditions

$$q(t_1) = q_1, \quad q(t_2) = q_2 \tag{5.2}$$

and the integral over $q(t)$ is an integral over all such paths of the exponential evaluated for each path. The measure of integration is not well defined a priori, but it is possible to give prescriptions that make sense. In classical mechanics, positions at two different times determine a path from the second order differential equation of motion, which yields a definite solution if two conditions are imposed. In contrast, quantum mechanics does not allow us to talk about trajectories. The path integral formulation manifestly refers to the myriad paths that may be taken by a particle in going from one point to another. It goes so far as to assign a probability amplitude to each path, and it is related to the classical action corresponding to that path. However, it is complex and it also involves a superposition, characteristic of quantum mechanics.

There is also a phase space form of the path integral

$$\int \int \mathcal{D}q(t)\mathcal{D}p(t) e^{\frac{i}{\hbar} \int_{t_1}^{t_2} dt[p(t)\dot{q}(t) - H(q(t),p(t))]}, \tag{5.3}$$

where the integration is over functions $q(t)$ with the given boundary conditions and over $p(t)$ without boundary conditions. From this form, one can obtain the coordinate space path integral by integrating over the momenta $p(t)$. This integral is a Gaussian one if the Hamiltonian involves the momenta quadratically. This form of the path integral can be derived from the operator formulation of quantum mechanics by dividing the interval between t_1, t_2 into small pieces and inserting intermediate states in the matrix element.

In the case of field theory, the analogue of the position q of a particle in quantum mechanics is a field $\phi(x)$. The analogue of the above path integral is an integral over field configurations:

$$\int \mathcal{D}\phi(x)e^{\frac{i}{\hbar}\int d^4x \mathcal{L}}, \tag{5.4}$$

where the exponential, evaluated for each field configuration, is integrated over all field configurations, boundary conditions being free. This represents a probability amplitude, but in field theory, one is more interested in Green functions. In this case, the relevant relation from quantum mechanics is

$$\langle q_2 | e^{-\frac{i}{\hbar}H(t_2-t_a)} q(t_a) e^{-\frac{i}{\hbar}H(t_a-t_b)} q(t_b) e^{-\frac{i}{\hbar}H(t_b-t_1)} | q_1 \rangle$$
$$\propto \int \mathcal{D}q(t) q(t_a) q(t_b) e^{\frac{i}{\hbar}\int_{t_1}^{t_2} dt\, L(q(t),\dot{q}(t))}, \tag{5.5}$$

where $t_2 > t_a > t_b > t_1$. This yields the following field theoretic relation:

$$\langle 0|T(\phi(x_1)\phi(x_2))|0\rangle \propto \int \mathcal{D}\phi\phi(x_1)\phi(x_2)e^{\frac{i}{\hbar}\int d^4x \mathcal{L}}. \tag{5.6}$$

The normalization can be fixed by using the normalization of the vacuum:

$$\langle 0|T(\phi(x_1)\phi(x_2))|0\rangle = \frac{\int \mathcal{D}\phi\phi(x_1)\phi(x_2)e^{\frac{i}{\hbar}\int d^4x \mathcal{L}}}{\int \mathcal{D}\phi e^{\frac{i}{\hbar}\int d^4x \mathcal{L}}}. \tag{5.7}$$

We have written this down for two field operators between vacuum states, but the generalization to an arbitrary product holds.

$$\langle 0|T(\phi(x_1)\phi(x_2)...\phi(x_n))|0\rangle = \frac{\int \mathcal{D}\phi\phi(x_1)\phi(x_2)...\phi(x_n)e^{\frac{i}{\hbar}\int d^4x \mathcal{L}}}{\int \mathcal{D}\phi e^{\frac{i}{\hbar}\int d^4x \mathcal{L}}}. \tag{5.8}$$

Here $\mathcal{L}$ is the Lagrangian density of the field theoretic model under consideration. A trick that can be used to combine such relations involves the addition of a hypothetical source. One defines

$$Z[j] = \int \mathcal{D}\phi e^{\frac{i}{\hbar}\int d^4x (\mathcal{L}+j(x)\phi(x))} \tag{5.9}$$

and notes that

$$\frac{\hbar}{i}\frac{\delta}{\delta j(x)}Z[j] = \int \mathcal{D}\phi\phi(x)e^{\frac{i}{\hbar}\int d^4x (\mathcal{L}+j\phi)}, \tag{5.10}$$

so that

$$\langle 0|T(\phi(x_1)\phi(x_2)...\phi(x_n))|0\rangle = (\frac{\hbar}{i})^n \frac{\delta}{\delta j(x_1)} \frac{\delta}{\delta j(x_2)} \cdots \frac{\delta}{\delta j(x_n)} \frac{Z[j]}{Z[0]}|_{j=0}. \quad (5.11)$$

This shows that all n-point functions of a field theoretic model can be calculated if the basic functional integral $Z[j]$ is known. The $j\phi$ term is linear in the fields and is not an extra complication if $Z[0]$ can be calculated. The Lagrangian density usually involves a free part, which is quadratic, together with an interaction part, which has cubic and higher degree terms in the fields. For a free field theory, only quadratic and linear terms occur in the exponent and the functional integrals can be evaluated exactly by analogy with Gaussian integrals.

The multidimensional Gaussian integral

$$\int \frac{dp_1}{\sqrt{\pi}} \int \frac{dp_1}{\sqrt{\pi}} \cdots \int \frac{dp_1}{\sqrt{\pi}} e^{-p^T M p + v^T p} = e^{\frac{1}{4} v^T M^{-1} v} (\det M)^{-\frac{1}{2}}, \quad (5.12)$$

where p, v are n-component column vectors and M an $n \times n$ matrix, suggests the definition

$$\int \mathcal{D}\phi e^{-\phi.M\phi+v.\phi} = e^{\frac{1}{4} v.M^{-1} v} (\det M)^{-\frac{1}{2}} \quad (5.13)$$

for the functional integral, with v now representing a function of x, like ϕ, M an operator acting on such functions and $v.\phi$ denoting $\int d^4 x v(x)\phi(x)$. The factors of $\sqrt{\pi}$ are not exhibited explicitly because of the ambiguity in the measure of the functional integration. Such factors cancel out in ratios of the $Z[j]$. The determinant of M may be interpreted as the product of its eigenvalues.

It is to be noted that the integrals mentioned here are of the Gaussian type, with exponential damping for large values of p or ϕ, while the intended application is to integrals with complex exponents without any such damping. Appropriate damping has to be provided by introducing suitable $i\epsilon$ terms as indicated below.

5.1 Scalar field theory

For the free scalar field theory,

$$\mathcal{L} = \frac{1}{2}(\partial_\mu \phi \partial^\mu \phi - m^2 \phi^2). \quad (5.14)$$

In this case, after setting $\hbar = 1$, one finds

$$
\begin{aligned}
Z[j] &= \int \mathcal{D}\phi \exp[i \int d^4x (\frac{1}{2}(\partial_\mu \phi \partial^\mu \phi - m^2\phi^2) + j\phi)] \\
&= \int \mathcal{D}\phi \exp[i \int d^4x (\frac{1}{2}(-\phi\Box\phi - \phi m^2\phi) + j.\phi)] \\
&= e^{-\frac{i}{2}\int\int d^4x d^4y j(x)\Delta_F(x-y)j(y)}[\det \frac{i(\Box + m^2)}{2}]^{-\frac{1}{2}}, \quad (5.15)
\end{aligned}
$$

where Δ is related to the inverse of the differential operator in the Lagrangian density by

$$
(\Box + m^2)\Delta_F(x - y) = -\delta^{(4)}(x - y). \quad (5.16)
$$

The exponential in the functional integral can be damped if the factor m^2 is replaced by $m^2 - i\epsilon$, which produces $-\frac{1}{2}\epsilon m^2\phi^2$. This shows up in the expression for Δ_F with $\epsilon \to 0$:

$$
\Delta_F(x - y) = \int \frac{d^4k}{(2\pi)^4} \frac{e^{-ik.(x-y)}}{k^2 - m^2 + i0}. \quad (5.17)
$$

The determinant in $Z[j]$ does not have to be calculated as it simply cancels out in all ratios. On the other hand, the inverse of the operator, which is essentially Δ_F, appears in Green functions. The propagator is

$$
\begin{aligned}
\langle 0|T(\phi(x)\phi(y))|0\rangle &= (\frac{1}{i})^2 \frac{\delta}{\delta j(x)} \frac{\delta}{\delta j(y)} \frac{Z[j]}{Z[0]}\Big|_{j=0} \\
&= i\Delta_F(x - y). \quad (5.18)
\end{aligned}
$$

This reproduces the expression known from canonical quantization of the free scalar field theory.

If interaction terms are present, they can be treated in perturbation theory. If

$$
\mathcal{L} = \mathcal{L}_0 + \mathcal{L}_I, \quad (5.19)
$$

where the first piece contains the free quadratic terms and the second piece contains higher degree terms, the exponential of the action can be factorized:

$$
\begin{aligned}
Z[j] &= \exp[i \int d^4x \mathcal{L}_I(\frac{1}{i}\frac{\delta}{\delta j})] \int \mathcal{D}\phi \exp[i \int d^4x (\mathcal{L}_0 + j\phi)] \\
&= \exp[i \int d^4x \mathcal{L}_I(\frac{1}{i}\frac{\delta}{\delta j})] Z_0[j], \quad (5.20)
\end{aligned}
$$

where $Z_0[j]$ refers to $Z[j]$ calculated for the free part of the Lagrangian density. $Z[j]$ can be evaluated in perturbation by developing the functional differential series produced by the exponential involving $\mathcal{L}_I$.

5.2 Fermion fields in functional formalism

Quantum field theoretic models in general may have scalar fields, vector fields, as well as spinor fields. While scalar fields can be handled by a straightforward extension to field theory of the functional integral formalism, spinor fields and gauge vector fields bring in extra complications. Gauge fields will be considered later. Our present concern is spinor fields.

As is well known from the operator formulation of spinor fields, there is a crucial difference between scalar and spinor field operators: the latter obey anticommutation relations rather than commutation relations which come from the generalization of the canonical commutation relations of quantum mechanics. Commutators in quantum mechanics and quantum field theory are proportional to Planck's constant, vanishing in the classical limit, meaning that the corresponding objects commute in the classical limit. On the other hand, spinor fields obey anticommutation relations, with anticommutators proportional to Planck's constant, signifying that any objects which correspond to them in the classical limit must anticommute with one another. Ordinary numbers do not anticommute, of course, but matrices may. Such objects have to be used in setting up functional integrals for spinor fields. These objects are called Grassmann variables. If ξ, η are Grassmann variables,

$$\xi\eta + \eta\xi = 0, \quad \xi^2 = 0 = \eta^2. \tag{5.21}$$

The second equation is merely the statement of anticommutation of ξ with itself. The vanishing of the square of a Grassmann variable means that higher powers too must vanish, so the most general polynomial is a linear one. Now functional integration needs integrals of these linear functions of a Grassmann variable to be defined. $\int d\xi$, being both a Grassmann variable and a constant – in a definite integral – can only vanish:

$$\int d\xi = 0. \tag{5.22}$$

However, $\int d\xi\, \xi$, which is again a constant, is not a Grassmann variable:

$$\int d\xi\, \xi\, \eta = -\int d\xi\, \eta\, \xi = \eta \int d\xi\, \xi, \tag{5.23}$$

as η anticommutes with both ξ and $d\xi$. Hence the integral may be a nonvanishing constant. This constant is chosen to be unity:

$$\int d\xi\, \xi = 1. \tag{5.24}$$

All integrals can then be defined from these two basic integrals by linearity because only linear functions have to be considered:

$$\int d\xi(a + b\xi) = b. \tag{5.25}$$

The double integral of an exponential of a product of Grassmann variables is of interest:

$$\int d\eta \int d\xi e^{a\xi\eta} = \int d\eta \int d\xi (1 + a\xi\eta) = a. \tag{5.26}$$

If there are two pairs of Grassmann variables,

$$
\begin{aligned}
\int d^2\eta \int d^2\xi e^{\xi_i a_{ij}\eta_j} &= \int d^2\eta \int d^2\xi (a_{11}a_{22}\xi_1\eta_1\xi_2\eta_2 + a_{12}a_{21}\xi_1\eta_2\xi_2\eta_1) \\
&= -\det a.
\end{aligned}
\tag{5.27}
$$

This generalizes to n pairs:

$$
\begin{aligned}
\int d^n\eta \int d^n\xi e^{\xi_i a_{ij}\eta_j} &= \int d^n\eta \int d^n\xi (a_{11}a_{22}...a_{nn}\xi_1\eta_1\xi_2\eta_2...\xi_n\eta_n + ...) \\
&= (-)^{n(n-1)/2}\det a.
\end{aligned}
\tag{5.28}
$$

The field theoretic analogue is

$$\int \mathcal{D}\eta \int \mathcal{D}\xi e^{\int d^4x \xi(x) A\eta(x)} = \det A, \tag{5.29}$$

where A is an operator and the determinant may be interpreted as the product of its eigenvalues. The sign is of course ambiguous, but constant factors cancel out in ratios of functional integrals and are of no relevance.

For a Dirac field, the propagator is given by

$$\langle 0|T(\psi_\alpha(x)\bar{\psi}_\beta(y))|0\rangle = \frac{\int \mathcal{D}\psi \int \mathcal{D}\bar{\psi}\psi_\alpha(x)\bar{\psi}_\beta(y)e^{i\int d^4z \bar{\psi}D\psi}}{\int \mathcal{D}\psi \int \mathcal{D}\bar{\psi}e^{i\int d^4z \bar{\psi}D\psi}}, \tag{5.30}$$

where the fields are now Grassmann fields, anticommuting with one another and D stands for the Dirac operator. Sources have to be introduced for $\psi, \bar{\psi}$:

$$
\begin{aligned}
Z[\eta, \bar{\eta}] &= \int \mathcal{D}\psi \int \mathcal{D}\bar{\psi}e^{i\int d^4x(\bar{\psi}D\psi + \bar{\eta}\psi + \bar{\psi}\eta)} \\
&= (\det D)e^{i\int d^4x \int d^4y \bar{\eta}(x)D^{-1}(x,y)\eta(y)},
\end{aligned}
\tag{5.31}
$$

where the inverse operator has been written out in its expanded form:

$$[D^{-1}\psi](x) = \int d^4y D^{-1}(x,y)\psi(y). \tag{5.32}$$

The propagator is calculated by differentiation with respect to the anticommuting sources and becomes

$$\langle 0|T(\psi_\alpha(x)\bar{\psi}_\beta(y))|0\rangle = iD^{-1}_{\alpha\beta}(x,y). \tag{5.33}$$

Chapter 6

Nonabelian gauge symmetry

6.1 Classical theory

Gauge symmetry usually refers to the invariance of a Lagrangian density under spacetime dependent field transformations. The most familiar gauge symmetry is the invariance of the electromagnetic action or Lagrangian density under transformations of the potential which leave the fields unchanged and hence the Lagrangian density, the action and the equations of motion too unaltered.

In this chapter we shall consider more complicated gauge potentials and gauge transformations. Let ϕ be a set of fields written as a column vector. One often has Lagrangian densities which exhibit invariance under transformations

$$\phi(x) \to U\phi(x), \qquad (6.1)$$

where U is one of a set of matrices, independent of spacetime coordinates. The set of matrices forms a symmetry group. As matrices do not commute in general, the group may be noncommutative or nonabelian. The simplest case is when the matrix has a single component and the group is abelian. This is what happens in the case of ordinary gauge symmetry in electrodynamics.

If this group is a continuous group with an infinite number of elements, one may try to extend the symmetry to spacetime dependent transformations, where U starts depending on x. This requires the Lagrangian density to have a special structure. For spacetime dependent transformations,

$$\partial_\mu\phi \to U\partial_\mu\phi + (\partial_\mu U)\phi, \qquad (6.2)$$

which is not exactly like the transformation of ϕ: there is an extra piece. As a result, a term like $\partial_\mu\phi^\dagger \partial^\mu\phi$, which is invariant under a spacetime independent transformation for unitary U, is not invariant under a spacetime dependent one. However, it is possible to modify it so as to make it invariant. This requires the introduction of a new field, similar to the electromagnetic potential, which can cancel the extra term occurring in the transformation of the field ϕ. Suppose

$$(\partial_\mu - igA_\mu)\phi \tag{6.3}$$

is the modified form of the partial derivative which transforms like ϕ. Note first that A_μ has to be a matrix which can multiply the column vector ϕ from the left. It has to transform in such a way that

$$(\partial_\mu - igA_\mu)\phi \rightarrow U(\partial_\mu - igA_\mu)\phi \tag{6.4}$$

or

$$(\partial_\mu - igA_\mu) \rightarrow U(\partial_\mu - igA_\mu)U^{-1}. \tag{6.5}$$

This requires that

$$(-igA_\mu)\phi \rightarrow U(-igA_\mu)\phi - (\partial_\mu U)\phi \tag{6.6}$$

or

$$A_\mu \rightarrow UA_\mu U^{-1} - \frac{i}{g}(\partial_\mu U)U^{-1}. \tag{6.7}$$

This transformation of A ensures that the modified partial derivative transforms in the same way as ϕ. The modified derivative is therefore called a covariant derivative: it varies in the same way as the undifferentiated field.

As is known from differential geometry, covariant differentiation requires the introduction of a connection. A_μ or the 1-form $A \equiv A_\mu dx^\mu$ provides the connection, which is matrix valued.

In the special case when ϕ has a single component, U too is a number. Calling it

$$U = e^{-ig\theta}, \tag{6.8}$$

we find

$$A_\mu \rightarrow A_\mu - \partial_\mu \theta, \tag{6.9}$$

which is the gauge transformation of the electromagnetic potential.

The above construction shows how a term exhibiting invariance under spacetime independent transformations can be extended to one having independence under spacetime dependent transformations by introducing a new field A_μ. Now we have to introduce a kinetic piece for this field. Consider for this purpose the curvature of the connection, which amounts to the commutator of two covariant derivative operators:

$$\begin{aligned}
[\partial_\mu - igA_\mu, \partial_\nu - igA_\nu] &\equiv -igF_{\mu\nu} \\
&= -ig(\partial_\mu A_\nu - \partial_\nu A_\mu - ig[A_\mu, A_\nu]). \tag{6.10}
\end{aligned}$$

In view of the covariant transformation of the covariant derivative operator, it follows that

$$F_{\mu\nu} \to U F_{\mu\nu} U^{-1}. \tag{6.11}$$

The commutator

$$[\partial_\mu - ig A_\mu, F_{\nu\rho}] = \partial_\mu F_{\nu\rho} - ig[A_\mu, F_{\nu\rho}] \tag{6.12}$$

also transforms in the same way and is the covariant derivative of F. It follows from the transformation of F that the trace of the matrix F or FF is invariant under gauge transformations. The natural generalization of the electromagnetic Lagrangian density is thus

$$-\frac{1}{4} \text{ tr } F_{\mu\nu} F^{\mu\nu} \text{ or } -\frac{1}{2} \text{ tr } F_{\mu\nu} F^{\mu\nu} \tag{6.13}$$

depending on the normalization of the basis matrices for the gauge fields. If the basis matrices are taken as T^a, satisfying

$$\text{tr } T^a T^b = \frac{1}{2}\delta^{ab}, \tag{6.14}$$

and the fields are expanded as

$$A_\mu = A_\mu^a T^a, \quad F_{\mu\nu} = F_{\mu\nu}^a T^a, \tag{6.15}$$

one has

$$-\frac{1}{2} \text{ tr } F_{\mu\nu} F^{\mu\nu} = -\frac{1}{4} F_{\mu\nu}^a F^{a\mu\nu}. \tag{6.16}$$

The basis matrices or generators of the Lie algebra will satisfy commutation relations like

$$[T^a, T^b] = i f^{abc} T^c, \tag{6.17}$$

where f^{abc}, the structure constants, are antisymmetric in the first two indices and can be arranged to be antisymmetric in all three. The definition of F yields

$$F_{\mu\nu}^a = \partial_\mu A_\nu^a - \partial_\nu A_\mu^a + g f^{abc} A_\mu^b A_\nu^c. \tag{6.18}$$

If for an infinitesimal transformation one sets

$$U = e^{-igT^a \theta^a}, \tag{6.19}$$

the gauge transformation of A_μ reads

$$A_\mu^a \to A_\mu^a - \partial_\mu \theta^a + g f^{abc} \theta^b A_\mu^c. \tag{6.20}$$

For the field $F_{\mu\nu}$, the transformation is homogeneous, as we have seen:

$$F^a_{\mu\nu} \to F^a_{\mu\nu} + g f^{abc} \theta^b F^c_{\mu\nu}. \tag{6.21}$$

One obtains from the Lagrangian density an equation of motion

$$\partial^\mu F^a_{\mu\nu} + g f^{abc} A^{b\mu} F^c_{\mu\nu} = -\frac{\partial \mathcal{L}_{int}}{\partial A^{a\nu}}, \tag{6.22}$$

where the right side contains the current interacting with the gauge field and the left side is the covariant derivative of the field tensor. The current is expected to transform covariantly in the same way as F or its covariant derivative. A Bianchi identity may be obtained by applying the Jacobi identity to the covariant derivative operator:

$$\partial_{[\mu} F^a_{\nu\lambda]} + g f^{abc} A^b_{[\mu} F^c_{\nu\lambda]} = 0. \tag{6.23}$$

The indices μ, ν, λ are antisymmetrized in this equation. While in the electrodynamic case these involve only F and not A, both are involved in the general situation.

6.2   Quantization

We have seen earlier that the electromagnetic theory leads to complications in quantization. There is a problem in finding the canonical conjugate of the scalar potential, which is related to gauge invariance. The quadratic part of the Lagrangian density is supposed to yield the propagator, but this cannot be defined for a gauge theory. The quadratic part is the same for the abelian and nonabelian cases:

$$\int d^4x (\partial_\mu A^a_\nu - \partial_\nu A^a_\mu)(\partial^\mu A^{a\nu} - \partial^\nu A^{a\mu}) = 2 \int d^4x$$
$$A^a_\mu (-g^{\mu\nu}\Box + \partial^\mu \partial^\nu) A^a_\nu. \tag{6.24}$$

In order to find the propagator, one needs the inverse of the differential operator $-g^{\mu\nu}\Box + \partial^\mu \partial^\nu$. However, the operator has zero modes:

$$(-g^{\mu\nu}\Box + \partial^\mu \partial^\nu)\partial_\nu \chi = 0 \tag{6.25}$$

for any field $\chi(x)$, implying that the differential operator is singular and cannot be inverted. This shows the connection between gauge invariance and the problem of quantization. As before, gauge fixing is done to carry out quantization. In the functional integral approach, one tends to sum over each gauge field configuration and its gauge transforms, all contributing equally because of gauge invariance. Thus the functional integral may be said to be proportional to the volume of the gauge group. But this volume must be infinite because the group has an infinite number of factors, one group attached to each spacetime point. This infinite volume has to be taken out as a factor. This can be done by fixing any gauge condition, say

$$f^a(A) = 0. \tag{6.26}$$

One then defines the Faddeev–Popov factor

$$\Delta_f(A) = \left(\int \mathcal{D}U \delta[f(A^U)] \right)^{-1}, \tag{6.27}$$

where A^U stands for the gauge transform of the configuration A by the gauge group element U and the delta function applies to each component of the gauge condition and each spacetime point. The integration on the right is over the gauge group, which has a gauge invariant measure. As a result, the integral is invariant under gauge transformations of A. Then we can write down the functional integral over gauge fields in the presence of sources. For convenience, we adopt a compact notation where $J_\mu^a A^{a\mu}$ is written as JA. Then,

$$
\begin{aligned}
Z[J] &= \int \mathcal{D}A e^{iS[A]+i\int JA} \Delta_f[A] \int \mathcal{D}U \delta[f(A^U)] \\
&= \int \mathcal{D}U \int \mathcal{D}A e^{iS[A]+i\int JA} \delta[f(A^U)] \Delta_f[A] \\
&= \int \mathcal{D}U \int \mathcal{D}A e^{iS[A]+i\int JA^{U-1}} \delta[f(A)] \Delta_f[A], \tag{6.28}
\end{aligned}
$$

where in the third step an inverse gauge transformation has been made, i.e., A has been replaced by that configuration which on gauge transformation by U leads to A. Use has been made of the gauge invariance of the action (without the source current) and Δ. The term involving J is not gauge invariant, but if gauge invariant objects are to be calculated, the error made by replacing a gauge transform of A by A will not show up. With this understanding, one finally has

$$Z[J] = \int \mathcal{D}U \int \mathcal{D}A e^{iS[A]+i\int JA} \delta[f(A)] \Delta_f[A]. \tag{6.29}$$

In this form of the functional integral, the gauge group integration has been separated out as a factor. The gauge fixing, implemented by the delta function of f, is expected to make the functional integral over A regular. What is noteworthy is that the process of removal of the gauge group volume has introduced a new factor $\Delta_f[A]$ into the functional integral.

Now by changing the integration variables from U to $f(A^U)$, one can write

$$
\begin{aligned}
\Delta_f(A) &= \left(\int \mathcal{D}U \delta[f(A^U)] \right)^{-1} \\
&= \left(\int \mathcal{D}f(A^U) \left(\frac{\delta f(A^U)}{\delta U} \right)^{-1} \delta[f(A^U)] \right)^{-1} \\
&= \left(\left(\frac{\delta f(A^U)}{\delta U} \right)^{-1} \Big|_{U=1} \right)^{-1} \\
&= \frac{\delta f(A^U)}{\delta U} \Big|_{U=1}.
\end{aligned}
\tag{6.30}
$$

Here the delta functional is used to simplify the functional integral, and U collapses to the identity transformation if A_μ is a configuration which satisfies the gauge condition $f(A) = 0$. It is assumed that there is no nontrivial gauge transformation of A which again makes it satisfy the gauge condition. If such transformations do exist, their contribution to the integral has to be taken into account. Gauge related solutions of gauge conditions are known as Gribov copies.

It is to be recalled here that the left side is gauge invariant. If A_μ does not satisfy the condition, then it has to be gauge transformed so that it satisfies the condition and U differs from the identity transformation, but the value of the left side is the same. The right side looks complicated, but it is clearly the Jacobian determinant of the transformation from the variables U to $f(A^U)$. In general, this is indeed complicated, but because of the requirement that U is the identity when the configuration satisfies the gauge condition, the Jacobian has to be evaluated only for infinitesimal transformations in the neighborhood of the identity. Then one can write

$$
f^a(A^U) \simeq f^a(A) + \int (M[A]\theta)^a + ...,
\tag{6.31}
$$

where near the identity, U can be represented by infinitesimal parameters θ^a. M is therefore a matrix. The functional derivative can be calculated with respect to θ in the neighborhood of vanishing θ, so that the determinant of M appears as Δ.

The phase space approach is instructive. The Lagrangian density $-\frac{1}{4} F^a_{\mu\nu} F^{a\mu\nu}$ leads to the primary constraints

$$
\pi^a_0 \equiv \frac{\partial \mathcal{L}}{\partial \dot{A}^{a0}} = 0
\tag{6.32}
$$

and the momenta

$$
\pi^a_i \equiv \frac{\partial \mathcal{L}}{\partial \dot{A}^{ai}} = F^a_{i0}.
\tag{6.33}
$$

The Hamiltonian density is

$$
\begin{aligned}
\mathcal{H} &= F_{i0}^a \dot{A}^{ia} - \mathcal{L} \\
&= F_{i0}^a (F^{0ia} + \partial^i A^{0a} - g f^{abc} A^{0b} A^{ic}) + \frac{1}{4} F_{\mu\nu}^a F^{\mu\nu a} \\
&= -F_{i0}^a F^{i0a} + \frac{1}{2} F_{i0}^a F^{i0a} + \frac{1}{4} F_{ij}^a F^{ija} + F_{i0}^a (\partial^i A^{0a} - g f^{abc} A^{0b} A^{ic}) \\
&= -\frac{1}{2} F_{i0}^a F^{i0a} + \frac{1}{4} F_{ij}^a F^{ija} - A^{0a} (\partial^i F_{i0}^a + g f^{abc} A^{ib} F_{i0}^c) + \partial^i (F_{i0}^a A^{0a}).
\end{aligned}
$$

$$(6.34)$$

The last term goes away when this is integrated to produce the Hamiltonian. The preceding term produces a secondary constraint when the Poisson bracket of the primary constraint and the Hamiltonian is calculated:

$$
G^a \equiv \{\pi_0^a, H\} = \partial^i F_{i0}^a + g f^{abc} A^{ib} F_{i0}^c = 0. \tag{6.35}
$$

This is the nonabelian version of Gauss' law. One can check that the constraints are first class. The Poisson brackets

$$
\{\pi_0^a, \pi_0^b\} = \{\pi_0^a, G^b\} = 0, \tag{6.36}
$$

while the Poisson brackets

$$
\{G^a(\vec{x}), G^b(\vec{y})\} = g f^{abc} G^c \delta(\vec{x} - \vec{y}) \tag{6.37}
$$

are nontrivial but vanish when the constraints are imposed. In fact, G^a is the generator of time independent gauge transformations:

$$
\{A_i^a(\vec{x}), G^b(\vec{y})\} = -\partial_i \delta(\vec{x} - \vec{y}) \delta^{ab} + g f^{abc} A_i^c \delta(\vec{x} - \vec{y}), \tag{6.38}
$$

which is equivalent to

$$
\{A_i^a, \int d^3 \vec{y} \theta^b G^b(\vec{y})\} = -\partial_i \theta^a + g f^{abc} \theta^b A_i^c. \tag{6.39}
$$

The time invariance of Gauss' law does not produce any new constraint:

$$
\{G^a, H\} = -g f^{abc} A_0^b G^c, \tag{6.40}
$$

which vanishes when the constraints are imposed.

To remove gauge degrees of freedom, gauge conditions are needed. The Gauss constraints need a conjugate set of gauge conditions $f^a(A) = 0$ which together are second class. These are supposed to form a pair of conjugate fields for each value of the index a, but it is not straightforward to solve these constraints to identify the unconstrained or physical fields and momenta explicitly. If one could do so, the phase space path integral would involve integration over only these physical phase space degrees of freedom. Instead, one integrates over all A_i^a, π_i^a, with delta functions imposing the Gauss constraints — already given by the functional integral over the Lagrange multiplier fields A_0^a because $\mathcal{H}$ in the exponent contains a product of A_0^a with G^a — *and* the gauge conditions f, together with a Jacobian factor for taking into account the fact that these constraints are *not* canonically conjugate fields. This factor is

$$\det \{f^a, G^b\} = \det M[A] = \Delta_f[A]. \qquad (6.41)$$

This is the origin of the Faddeev–Popov determinant factor in the canonical approach.

Recalling that a determinant is produced by an integral over Grassmann variables, one can use a representation for the Faddeev–Popov determinant as

$$\Delta_f[A] = \det M[A] \propto \int \int \mathcal{D}c \mathcal{D}\bar{c} e^{i \int \int \bar{c}Mc}. \qquad (6.42)$$

Hence, after dropping the gauge volume factor, the functional integral can be written as

$$Z[J] = \int \int \int \mathcal{D}A \mathcal{D}c \mathcal{D}\bar{c} e^{iS[A]+i \int JA+i \int \int \bar{c}Mc} \delta[f(A)]. \qquad (6.43)$$

This is how hypothetical fermions, called ghosts, enter the functional integral formulation of gauge theories. It is clear that the c fields too carry the index a and the matrix M carries a pair of indices. The delta functions too can be exponentiated by introducing auxiliary Lagrange multiplier fields carrying similar indices:

$$Z[J] = \int \int \int \int \mathcal{D}A \mathcal{D}B \mathcal{D}c \mathcal{D}\bar{c} e^{iS[A]+i \int JA+i \int Bf(A)+i \int \int \bar{c}Mc}. \qquad (6.44)$$

Now let us consider the Lorentz scalar gauge fixing conditions

$$\partial_\mu A^{a\mu} = 0. \qquad (6.45)$$

This transforms to

$$\partial_\mu A^{a\mu} - \Box\theta^a + gf^{abc}\partial_\mu(\theta^b A^{c\mu}). \qquad (6.46)$$

Hence we identify in this particular case

$$M[A] = -\delta^{ab}\Box + gf^{abc}\partial_\mu A^{c\mu} + gf^{abc}A^{c\mu}\partial_\mu. \qquad (6.47)$$

Consequently, the ghost Lagrangian density for this gauge condition is

$$\bar{c}^a \partial_\mu(-\partial^\mu c^a + gf^{abc}c^b A^{c\mu}). \qquad (6.48)$$

Equivalently, one can replace it through integration by parts by

$$\partial_\mu \bar{c}^a(\partial^\mu c^a - gf^{abc}c^b A^{c\mu}). \qquad (6.49)$$

The ghost is a scalar field with an internal index which couples to the gauge field and is quantized as a fermion in spite of its spinlessness. It does not appear as a physical particle, of course. Note that it decouples if the gauge coupling constant vanishes and also in the case of an abelian gauge field.

6.3   BRS symmetry

The classical Lagrangian density of the gauge field, including interactions with other fields, is invariant under spacetime dependent gauge transformations. When a gauge fixing condition is used, this symmetry is broken. Spacetime independent gauge transformations continue to be a symmetry of the Lagrangian density. But there is a very interesting new symmetry involving the ghosts as well as the gauge fields which is of great help in many calculations. This is the Becchi–Rouet–Stora or BRS symmetry, involving a special gauge-like transformation.

First consider the transformation of A^μ, which resembles an infinitesimal gauge transformation, but with a parameter involving the ghost field c so that the gauge field Lagrangian is invariant:

$$\delta A^{a\mu} = \partial^\mu c^a + g f^{abc} A^{b\mu} c^c. \tag{6.50}$$

The transformation of c is given by

$$\delta c^a = -\frac{1}{2} g f^{abc} c^b c^c. \tag{6.51}$$

The right sides have one higher Grassmann power than the left sides, meaning that δ itself is of a Grassmann nature. It has to be taken to anticommute with c. With this understanding, we see that

$$\delta^2 c^a = \frac{1}{4} g^2 (f^{abc} f^{bde} c^d c^e c^c - f^{abc} f^{cde} c^b c^d c^e) = \frac{1}{2} g^2 f^{abc} f^{bde} c^c c^d c^e, \tag{6.52}$$

where the anticommutation of δ and c has been used in the first equality and the antisymmetry of f in the next. Now using the antisymmetry of the product $c^c c^d c^e$ in the three indices and the Jacobi identity satisfied by the structural constants, we conclude that

$$\delta^2 c^a = 0, \tag{6.53}$$

as is expected for a Grassmann δ. Next we consider

$$\delta^2 A^{a\mu} = -\frac{1}{2} g f^{abc} \partial^\mu (c^b c^c) + g f^{abc} (\partial^\mu c^b + g f^{bde} A^{d\mu} c^e) c^c$$
$$-\frac{1}{2} g f^{abc} f^{cde} A^{b\mu} c^d c^e. \tag{6.54}$$

The quadratic terms read

$$
\begin{aligned}
-\frac{1}{2} g f^{abc} \partial^\mu (c^b c^c) + g f^{abc} \partial^\mu c^b c^c &= -\frac{1}{2} g f^{abc} \partial^\mu (c^b c^c) - g f^{abc} \partial^\mu c^c c^b \\
&= -\frac{1}{2} g f^{abc} \partial^\mu (c^b c^c) + g f^{abc} c^b \partial^\mu c^c \\
&= -\frac{1}{2} g f^{abc} \partial^\mu (c^b c^c) + \frac{1}{2} g f^{abc} \partial^\mu (c^b c^c) \\
&= 0. \tag{6.55}
\end{aligned}
$$

Here, in the third line two forms of the second term have been averaged over. The cubic terms can now be collected:

$$g^2 f^{abc} f^{bde} A^{d\mu} c^e c^c - \frac{1}{2} g^2 f^{abc} f^{cde} A^{b\mu} c^d c^e$$

$$= g^2 f^{abc} f^{bde} A^{d\mu} c^e c^c - \frac{1}{2} g^2 f^{adb} f^{bec} A^{d\mu} c^e c^c. \qquad (6.56)$$

Using the antisymmetry in the indices in $c^e c^c$, one can now write

$$(f^{abc} f^{bde} - \frac{1}{2} f^{adb} f^{bec}) c^e c^c = \frac{1}{2}(-f^{acb} f^{bde} - f^{aeb} f^{bcd} - f^{adb} f^{bec}) c^e c^c$$

$$= 0, \qquad (6.57)$$

which follows from the Jacobi identity of the structure constants. Hence

$$\delta^2 A^{a\mu} = 0. \qquad (6.58)$$

This property is thus verified on both c and A and is consistent with the anticommuting nature of δ.

Next let us consider the transformation of the gauge fixing and ghost terms in the Lagrangian density. For the particular gauge condition introduced in the previous section,

$$\delta[-\bar{c}^a \partial_\mu (\partial^\mu c^a - g f^{abc} A^{c\mu} c^b)] = \delta[-\bar{c}^a \partial_\mu \delta A^{a\mu}] = -[\delta \bar{c}^a] \partial_\mu \delta A^{a\mu} \qquad (6.59)$$

because $\delta^2 A^{a\mu}$ vanishes. Note that we have not yet defined the transformation of $\bar{c}$. More generally, for a gauge fixing condition $f(A) = 0$,

$$\delta[\bar{c}^a M^{ab} c^b] = -\delta[\bar{c}^a \delta f^a] = -[\delta \bar{c}^a] \delta f^a. \qquad (6.60)$$

Now the transformation of the gauge fixing piece is

$$\delta[B^a f^a]. \qquad (6.61)$$

The above two pieces cancel each other provided one sets

$$\delta \bar{c}^a = B^a, \qquad (6.62)$$

and

$$\delta B^a = 0, \qquad (6.63)$$

which also ensure that

$$\delta^2 \bar{c}^a = 0, \qquad (6.64)$$

as is required by the Grassmann nature of δ.

In the case of c, A, the degree of anticommuting Grassmann variables increases by 1 under the action of δ, while $\delta\bar{c}$ has degree zero in Grassmann variables. But δc is also of degree zero, modulo 2. A degree 2 makes an object Grassmann even, i.e., it commutes with Grassmann variables instead of anticommuting with them.

The original Lagrangian density before gauge fixing as also the gauge fixing and ghost terms together are thus gauge invariant. Now fields coupling to the gauge field, which transform under a gauge transformation, have also to be transformed under δ. We write

$$\delta\phi = igT^a c^a \phi, \tag{6.65}$$

so that

$$\delta^2\phi = igT^a(-\frac{1}{2}gf^{abc}c^b c^c)\phi - (ig)^2 T^a T^b c^a c^b \phi = 0, \tag{6.66}$$

as can be seen by replacing the product of the two T's in the second term by half its commutator in view of the antisymmetry in the indices of the anticommuting ghosts. This confirms the nilpotency of δ:

$$\delta^2 = 0. \tag{6.67}$$

The interaction terms have to be invariant because the transformation has the structure of a gauge transformation. Hence the full Lagrangian density is invariant under the BRS transformation δ.

Chapter 7

Discrete symmetries

We have discussed rotations and Lorentz transformations in an earlier chapter. These are continuous spacetime symmetries shown by classical and quantum field theories. They give rise to conserved charges with a continuous spectrum of values. But there are also a few discrete symmetries, some associated with spacetime, which are seen in many systems: parity or space inversion and time reversal, as also charge conjugation, which has nothing to do with spacetime. Parity corresponds to $\vec{x} \to -\vec{x}$, while time reversal corresponds to $x^0 \to -x^0$. One considers the behavior of the action under field transformations

$$\phi(x^0, \vec{x}) \to \phi(x^0, -\vec{x}) \tag{7.1}$$

and

$$\phi(x^0, \vec{x}) \to \phi(-x^0, \vec{x}). \tag{7.2}$$

It is clear that the free scalar field theory action is invariant under these transformations. One can also insert ± 1 in the right sides of these equations without spoiling the invariance. The sign becomes important in interacting theories. In the case of parity, the factor ± 1 is called the intrinsic parity of the field. The $+1$ factor corresponds to a scalar and the -1 factor to a pseudoscalar, which changes sign under a reflection.

The case of the electromagnetic field is slightly more complicated. This is because of the occurrence of four components A_μ in the Lagrangian density

$$-\frac{1}{4}(\partial_\mu A_\nu - \partial_\nu A_\mu)(\partial^\mu A^\nu - \partial^\nu A^\mu). \tag{7.3}$$

For parity invariance, the spatial components have to possess the same intrinsic parity and the time component the opposite value. One normally takes A_0 to be a scalar (rather than a pseudoscalar) and then $\vec{A}$ has to be a vector rather than an axial vector. This is in accord with Maxwell's equations, which involve interactions in addition to the free theory.

The discrete transformations of the fields are implemented by unitary or – in the case of time reversal – antiunitary operators. In theories where these transformations are symmetries, there are discrete conservation laws – corresponding to parity, time reversal or charge conjugation *quantum numbers.* These are multiplicative quantum numbers.

We shall discuss the transformations for spinor fields in detail.

7.1 Parity in spinor field theory

The parity invariance of a Dirac field theory means that there should be a transformation

$$\mathcal{L}(\vec{x}) \to \mathcal{P}\mathcal{L}(\vec{x})\mathcal{P}^{-1} = \mathcal{L}(-\vec{x}). \tag{7.4}$$

The time dependence of the fields has been suppressed for simplicity. The fields themselves have to transform as

$$\psi(\vec{x}) \to \mathcal{P}\psi(\vec{x})\mathcal{P}^{-1} = P\psi(-\vec{x}), \tag{7.5}$$

where P may be a matrix in spinor space. The free Dirac Lagrangian density

$$\mathcal{L} = \bar{\psi}(i\gamma^\mu \partial_\mu - m)\psi \tag{7.6}$$

satisfies the above equation provided the time derivative piece, the space derivative piece and the mass term transform correctly. This requires that

$$\begin{aligned} P^\dagger P &= 1, \\ -P^\dagger \gamma^0 \gamma^i P &= \gamma^0 \gamma^i, \\ P^\dagger \gamma^0 P &= \gamma^0. \end{aligned} \tag{7.7}$$

The second of these equations can be written as

$$-P^\dagger \gamma^0 P P^\dagger \gamma^i P = \gamma^0 \gamma^i, \tag{7.8}$$

which, with the help of the third equation, leads to

$$-P^\dagger \gamma^i P = \gamma^i. \tag{7.9}$$

All these are satisfied by $P = \gamma^0$. This is a matrix, making this case somewhat nontrivial. Additional phase factors can be introduced in P without spoiling the conditions given above. It is to be noted that the first condition, which ensures that the time derivative piece transforms correctly, also guarantees that the equal time anticommutation relations are unaffected.

Note that the fact that the mass term of the action remains invariant, as imposed in the above analysis, means that the combination $\bar{\psi}\psi$ is a scalar:

$$\bar{\psi}(\vec{x})\psi(\vec{x}) \to \bar{\psi}(-\vec{x})\psi(-\vec{x}). \tag{7.10}$$

The anticommutation of $P = \gamma^0$ with γ^5 implies that

$$\bar{\psi}(\vec{x})\gamma^5\psi(\vec{x}) \rightarrow -\bar{\psi}(-\vec{x})\gamma^5\psi(-\vec{x}), \tag{7.11}$$

indicating that $\bar{\psi}\gamma^5\psi$ is a pseudoscalar. Hence an interaction $\bar{\psi}\gamma^5\phi\psi$ will satisfy parity invariance if ϕ transforms like a pseudoscalar, but not if it transforms like a scalar. On the other hand, $\bar{\psi}\phi\psi$ will conserve parity if ϕ transforms like a scalar, but not if it transforms like a pseudoscalar. In the case of an interaction $\bar{\psi}\frac{1-\gamma^5}{2}\phi\psi$, where a chiral combination interacts with ϕ,

$$\bar{\psi}(\vec{x})\frac{1-\gamma^5}{2}\psi(\vec{x}) \rightarrow \bar{\psi}(-\vec{x})\frac{1+\gamma^5}{2}\psi(-\vec{x}). \tag{7.12}$$

The hermitian combination breaks parity for a charged scalar field ϕ:

$$\bar{\psi}\frac{1-\gamma^5}{2}\phi\psi(\vec{x}) + \bar{\psi}\frac{1+\gamma^5}{2}\phi^\dagger\psi(\vec{x})$$
$$\rightarrow \quad \bar{\psi}\frac{1+\gamma^5}{2}\phi\psi(-\vec{x}) + \bar{\psi}\frac{1-\gamma^5}{2}\phi^\dagger\psi(-\vec{x}). \tag{7.13}$$

What happens when gauge interactions are introduced into the free theory? The covariant derivative operator appearing in the action involves the combination $\partial_\mu - igA_\mu$. The above conditions ensure that the $\mu = 0$ part transforms correctly if A_0 is transformed like a scalar. The vector part transforms correctly if $\vec{A}$ is transformed like a vector. So parity is conserved in this situation.

Gauge fields may couple in other ways to spinor fields. The above discussion indicates that $\bar{\psi}\gamma^\mu\psi$, which entered the interaction, transforms like a vector, but one can see that $\bar{\psi}\gamma^\mu\gamma^5\psi$ will transform like an axial vector because of the extra γ^5. If this combination interacts with a gauge vector field, parity will obviously be violated. Weak interactions involve the coupling of chiral currents to gauge vector fields. This means that a combination

$$\bar{\psi}\gamma^\mu\frac{1-\gamma^5}{2}A_\mu\psi \tag{7.14}$$

describes the interaction. The vector part of the interaction conserves parity, but the axial current produces a pseudoscalar when contracted with the vector gauge field and this breaks parity. Thus weak interactions break parity in an essential way.

It is thus clear that a sum of a vector and an axial vector or a sum of a scalar and a pseudoscalar has a mixed transformation. However, this has led to some confusion in the literature. What happens if the mass term, which normally looks like $\bar{\psi}\psi$, is replaced by a superposition of this scalar combination with a pseudoscalar combination? One might expect parity to be violated, but since it is a free part of the action that is altered here, rather than an interaction piece, it is necessary to consider the issue afresh. Let us consider the Lagrangian density

$$\mathcal{L} = \bar{\psi}(i\gamma^\mu\partial_\mu - me^{i\gamma^5\theta})\psi \tag{7.15}$$

and attempt to ensure the invariance of the time derivative piece, the space derivative piece and the new mass term of the action, following the logic of our earlier analysis. There are three equations, of which the first two are unaltered, but the third one contains the new factor:

$$
\begin{aligned}
P^\dagger P &= 1, \\
-P^\dagger \gamma^0 \gamma^i P &= \gamma^0 \gamma^i, \\
P^\dagger \gamma^0 e^{i\gamma^5\theta} P &= \gamma^0 e^{i\gamma^5\theta}.
\end{aligned}
\tag{7.16}
$$

The second equation is now written in the form

$$
-P^\dagger \gamma^0 e^{i\gamma^5\theta} P P^\dagger \gamma^i e^{i\gamma^5\theta} P = \gamma^0 e^{i\gamma^5\theta} \gamma^i e^{i\gamma^5\theta},
\tag{7.17}
$$

which is justified by the fact that $e^{i\gamma^5\theta} \gamma^\mu e^{i\gamma^5\theta} = \gamma^\mu$. In view of the third equation of the trio, we can now infer that

$$
-P^\dagger \gamma^i e^{i\gamma^5\theta} P = \gamma^i e^{i\gamma^5\theta},
\tag{7.18}
$$

which, together with that third equation, can be satisfied by

$$
P = \gamma^0 e^{i\gamma^5\theta} = e^{-i\gamma^5\theta/2} \gamma^0 e^{i\gamma^5\theta/2}.
\tag{7.19}
$$

This means that if the mass term is modified by such a phase factor involving γ^5, the parity matrix P itself is changed, and the action becomes invariant under this redefined parity operation on ψ. It is interesting to note that under this parity transformation,

$$
\bar{\psi} e^{i\gamma^5\theta} \psi(\vec{x}) \rightarrow \bar{\psi} e^{i\gamma^5\theta} \psi(-\vec{x}),
\tag{7.20}
$$

so that the mass term, which looks like a mixture of a scalar and a pseudoscalar under the usual parity transformation, is actually a scalar under the redefined parity transformation.

Such a mass term can be generated by spontaneous symmetry breaking from a chiral interaction with a scalar field. Although such terms violate parity in the presence of a mass term, the chiral interaction term, if present *without an explicit mass term*, will show parity invariance with P altered as above.

Another term that is often invoked in nonabelian gauge field theory is

$$
\epsilon^{\mu\nu\rho\sigma} \operatorname{tr} F_{\mu\nu} F_{\rho\sigma}.
\tag{7.21}
$$

The term is a total derivative and has no effect in the classical theory, but may have a topological effect in the quantum theory. Because of the presence of the ϵ tensor, all indices have to be different. So there is one time index and three space indices in each term. Consequently, the term is parity odd, i.e., violates parity.

7.2   Time reversal in spinor field theory

The time reversal invariance of a fermionic action amounts to

$$\mathcal{T}\mathcal{L}(t)\mathcal{T}^{-1} = \mathcal{L}(-t). \tag{7.22}$$

The space dependence of the fields has been suppressed for simplicity. The field is taken to transform as

$$\mathcal{T}\psi(t)\mathcal{T}^{-1} = T\psi(-t) \tag{7.23}$$

with a suitable matrix T. If there exists a T preserving the invariance of the action, one has time reversal invariance, otherwise it is broken.

Recall first that time reversal has to be an antilinear operation. With the usual fermion mass term, we consider $\mathcal{L} = \bar{\psi}(i\gamma^\mu\partial_\mu - m)\psi$. Unless $i \to -i$, it is not possible to produce a negative sign in the time derivative piece. Assuming this antilinearity, one finds that for time reversal invariance, the matrices T have to obey

$$\begin{aligned}
T^\dagger T &= 1, \\
-T^\dagger \gamma^{0*}\gamma^{i*}T &= \gamma^0\gamma^i, \\
T^\dagger \gamma^{0*}T &= \gamma^0,
\end{aligned} \tag{7.24}$$

if the time derivative piece, the space derivative piece and the mass term are to transform correctly. Now the second of these equations can be rewritten as

$$-T^\dagger \gamma^{0*}TT^\dagger\gamma^{i*}T = \gamma^0\gamma^i, \tag{7.25}$$

and with the help of the third equation, this becomes

$$-T^\dagger \gamma^{i*}T = \gamma^i. \tag{7.26}$$

In the standard representation of gamma matrices, γ^2 is purely imaginary, while the rest are real. So T has to commute with γ^0, γ^2 and anticommute with γ^1, γ^3. Thus one can take

$$T = i\gamma^1\gamma^3 \tag{7.27}$$

up to a phase. It is to be noted that the first condition, which ensures that the time derivative piece transforms correctly, also guarantees that the equal time anticommutation relations are unaffected.

For a mass term with a phase involving γ^5, the situation changes: the Lagrangian density $\mathcal{L} = \bar{\psi}(i\gamma^\mu\partial_\mu - me^{i\theta\gamma_5})\psi$ leads to the modified requirement

$$\begin{aligned}
T^\dagger T &= 1, \\
-T^\dagger \gamma^{0*}\gamma^{i*}T &= \gamma^0\gamma^i, \\
T^\dagger \gamma^{0*}e^{-i\theta\gamma_5^*}T &= \gamma^0 e^{i\theta\gamma_5}.
\end{aligned} \tag{7.28}$$

Now the second of these equations can be rewritten as

$$-T^\dagger \gamma^{0*} e^{-i\theta\gamma_5^*} T T^\dagger \gamma^{i*} e^{-i\theta\gamma_5^*} T = \gamma^0 e^{i\theta\gamma_5} \gamma^i e^{i\theta\gamma_5}, \qquad (7.29)$$

and with the help of the third equation, this becomes

$$-T^\dagger \gamma^{i*} e^{-i\theta\gamma_5^*} T = \gamma^i e^{i\theta\gamma_5}. \qquad (7.30)$$

These are satisfied in the standard representation by

$$T = i e^{i\theta\gamma_5} \gamma^1 \gamma^3. \qquad (7.31)$$

Thus, as in the case of parity, time reversal invariance too works out with a modified matrix T.

Gauge field interactions have to be considered next. Under the antilinear transformation $\mathcal{T}$, the gauge fields have to transform differently from ∂_μ because of i:

$$\begin{aligned} A_0(t) &\rightarrow A_0(-t), \\ A_i(t) &\rightarrow -A_i(-t). \end{aligned} \qquad (7.32)$$

These are consistent with electrodynamics, where the vector potential is odd under time reversal because of its association with a current. These ensure the time reversal invariance of gauge field interactions with any Dirac field. In the case of chiral or axial vector currents too, time reversal invariance is not broken because T commutes with γ^5 and therefore vector and axial vector currents have similar time reversal transformations. This is unlike the situation with parity.

The scalar $\bar\psi\psi$ transforms to itself at the reversed time and so does the pseudoscalar $\bar\psi\gamma^5\psi$. Interactions involving any combination of these with a scalar field ϕ simply transform in the appropriate way. For the chiral combination, the hermitian sum

$$\bar\psi \frac{1-\gamma^5}{2} \phi\psi + \bar\psi \frac{1+\gamma^5}{2} \phi^\dagger\psi \qquad (7.33)$$

conserves time reversal. However, because of antilinearity, a complex coupling constant, if present here, can violate time reversal:

$$\begin{aligned} g\bar\psi \tfrac{1-\gamma^5}{2} \phi\psi(t) + g^*\bar\psi \tfrac{1+\gamma^5}{2} \phi^\dagger\psi(t) \\ \rightarrow g^*\bar\psi \tfrac{1-\gamma^5}{2} \phi\psi(-t) + g\bar\psi \tfrac{1+\gamma^5}{2} \phi^\dagger\psi(-t). \end{aligned} \qquad (7.34)$$

However, such chiral interactions can generate mass by spontaneous symmetry breaking. If the entire fermion mass is generated by this method and there is no explicit mass term in the Lagrangian density, T is altered as discussed above and time reversal invariance is preserved by the mass term.

What about the nonabelian gauge field term (7.21)? There is one time index among the four indices, making the term odd under time reversal too.

7.3 Charge conjugation

Charge conjugation is another discrete transformation that roughly can be thought of as a particle-antiparticle transformation. It is thus not associated with spacetime transformations. For a free scalar field, which corresponds to what may be regarded as a neutral particle, this transformation is trivial:

$$\phi \to \phi, \tag{7.35}$$

but for a charged scalar field, which may be regarded as a combination of two neutral scalar fields $\frac{1}{\sqrt{2}}(\phi_1 + i\phi_2)$,

$$\phi \to \phi^\dagger, \tag{7.36}$$

or equivalently,

$$\phi_1 \to \phi_1, \quad \phi_2 \to -\phi_2. \tag{7.37}$$

The free field action is invariant under these transformations.

It is more interesting when the charged scalar field is coupled to the electromagnetic field. For

$$(\partial_\mu - ieA_\mu)\phi \tag{7.38}$$

to transform to

$$(\partial_\mu + ieA_\mu)\phi^\dagger, \tag{7.39}$$

it is necessary that

$$A_\mu \to -A_\mu. \tag{7.40}$$

A Dirac field couples to the electromagnetic field through the vector current $\bar\psi\gamma^\mu\psi$, so under charge conjugation, one expects

$$\bar\psi\gamma^\mu\psi \to -\bar\psi\gamma^\mu\psi. \tag{7.41}$$

If one wants the charge conjugation operation to be linear and to be independent of spacetime, this sign is apparently in conflict with the invariance of the Lagrangian density, which involves essentially the same factors, apart from derivatives. The conflict can be resolved by integration by parts in the action. Just as in the case of charged scalar fields, where the field is conjugated by the charge conjugation operation, ψ has to be replaced by $\psi^\dagger$. But ψ has four components, so one has to be careful with them. Instead of $\psi^\dagger$, which is a row vector, one writes ψ^*, which is a column like ψ, but conjugate to it. It can also be thought of as

$$\psi^* \equiv (\psi^\dagger)^T. \tag{7.42}$$

The transformation

$$\psi \to C\psi^*, \tag{7.43}$$

where C is a matrix in spinor space, will lead to the above-mentioned transformation of the current if the components of the fields are taken to anticommute with one another and C can be chosen to satisfy

$$C^\dagger \gamma^0 \gamma^\mu C = (\gamma^0 \gamma^\mu)^*, \tag{7.44}$$

which entails

$$C^\dagger C = 1 \tag{7.45}$$

and

$$C^\dagger \gamma^0 \gamma^i C = (\gamma^0 \gamma^i)^*. \tag{7.46}$$

One will then have

$$\bar{\psi}\gamma^\mu \partial_\mu \psi \to -(\partial_\mu \bar{\psi})\gamma^\mu \psi, \tag{7.47}$$

leading to the correct transformation for the kinetic term in the action on integration by parts. To find out whether such a C exists, it is necessary to investigate what happens to the mass term. The usual mass term

$$\bar{\psi}\psi \to -\bar{\psi}\gamma^0 C^T \gamma^{0T} C^* \psi, \tag{7.48}$$

which reproduces the mass term provided

$$C^\dagger \gamma^0 C = -\gamma^{0*}. \tag{7.49}$$

This means that one needs, for all μ,

$$C^\dagger \gamma^\mu C = -\gamma^{\mu*}. \tag{7.50}$$

In the representation where γ^2 is pure imaginary and the others are real, C has to commute with γ^2 and to anticommute with γ^0, γ^1 and γ^3. A solution up to phase factors is

$$C = i\gamma^2. \tag{7.51}$$

In the representation where all four γ matrices are pure imaginary, C has to commute with all four and is given up to a phase by

$$C' = 1. \tag{7.52}$$

There are other ways of defining the transformation of ψ under charge conjugation with other forms of the matrix C.

It may be noted that the unitarity of C ensures that the equal time anti-commutation relations are preserved.

If the mass term is altered to $\bar{\psi}e^{i\gamma^5\theta}\psi$, the last requirement is modified to

$$C^\dagger\gamma^0 e^{i\theta\gamma^5}C = -(\gamma^0 e^{i\theta\gamma^5})^*, \tag{7.53}$$

which agrees with the equations satisfied by C because γ^5 contains an i and an even number of γ matrices. Thus, the charge conjugation matrix does not change if the mass term involves a phase with γ^5. This is different from what happens with P and T.

Similarly, the nonabelian gauge field term (7.21) does not change under charge conjugation. Because of odd parity, it is odd under CP, just as it is odd under time reversal for a real coupling constant, so that it is even under CPT.

We have seen that vector current interactions with gauge fields are invariant under charge conjugation. What about axial current interactions? The condition for C invariance of such terms is obtained from that for vector currents by introducing γ^5:

$$C^\dagger\gamma^0\gamma^\mu\gamma^5 C \overset{?}{=} (\gamma^0\gamma^\mu\gamma^5)^*. \tag{7.54}$$

As γ^5 contains an i and an even number of γ matrices, this differs in sign from the earlier condition. Hence axial interactions involving γ^5 break C just as they break P. However, the sign factors for C and P cancel out, so that CP is not violated by axial or chiral interactions with gauge fields. This is especially relevant for weak interactions, which break parity but not the combination CP.

As the scalar and pseudoscalar combinations of spinor fields simply transform to themselves under charge conjugation, interactions with charged scalar fields violate C. In particular,

$$\bar{\psi}\frac{1-\gamma^5}{2}\phi\psi + \bar{\psi}\frac{1+\gamma^5}{2}\phi^\dagger\psi \to \bar{\psi}\frac{1-\gamma^5}{2}\phi^\dagger\psi + \bar{\psi}\frac{1+\gamma^5}{2}\phi\psi. \tag{7.55}$$

As parity transforms the combination in the complementary way, CP is conserved here. Of course, if there is a complex coupling constant, the two terms have different coefficients, so that under CP

$$g\bar{\psi}\frac{1-\gamma^5}{2}\phi\psi(\vec{x}) + g^*\bar{\psi}\frac{1+\gamma^5}{2}\phi^\dagger\psi(\vec{x})$$
$$\to g\bar{\psi}\frac{1+\gamma^5}{2}\phi^\dagger\psi(-\vec{x}) + g^*\bar{\psi}\frac{1-\gamma^5}{2}\phi\psi(-\vec{x}), \tag{7.56}$$

signifying CP violation. This behavior is complementary to that under time reversal, so that the combination CPT yields

$$g\bar{\psi}\frac{1-\gamma^5}{2}\phi\psi(x) + g^*\bar{\psi}\frac{1+\gamma^5}{2}\phi^\dagger\psi(x)$$
$$\to g^*\bar{\psi}\frac{1+\gamma^5}{2}\phi^\dagger\psi(-x) + g\bar{\psi}\frac{1-\gamma^5}{2}\phi\psi(-x), \tag{7.57}$$

TABLE 7.1:   Behavior of
Interaction Terms under Discrete
Symmetries.

Term	C	P	T
$\bar{\psi}\gamma^{\mu}A_{\mu}\psi$	even	even	even
$\bar{\psi}\gamma^{5}\gamma^{\mu}A_{\mu}\psi$	odd	odd	even
$\bar{\psi}\phi\psi$ + h.c.	even	even	even
$\bar{\psi}\gamma^{5}\phi\psi$ + h.c.	odd	odd	even
$\epsilon^{\mu\nu\rho\sigma}F_{\mu\nu}F_{\rho\sigma}$	even	odd	odd

signifying CPT conservation. CPT can be violated only if the Lagrangian density can be nonhermitian, with nonconjugate coefficients for the two terms. A complex gauge coupling constant, producing a nonhermitian term, would similarly violate CPT besides T.

The charge conjugation transformation is useful in constructing some special products. $C\psi^{*}$ has the same Lorentz transformation as ψ, because

$$C\sigma^{\mu\nu*} = -\sigma^{\mu\nu}C, \tag{7.58}$$

which follows from the construction of C. Hence one can use a mass term

$$\bar{\psi}C\psi^{*} + \text{hermitian conjugate} \tag{7.59}$$

as a special mass term, called a Majorana mass term. This may be rewritten as

$$\psi^{*T}\gamma^{0}C\psi^{*} + \text{hermitian conjugate.} \tag{7.60}$$

Now the matrix $\gamma^{0}C$ is antisymmetric, so the term may appear to vanish identically, but does not do so because the components of ψ^{*} anticommute with one another. It is even under charge conjugation, but odd under parity and time reversal with the usual P, T. This mass term is not invariant even under a global phase transformation of the fermion field.

To conclude the discussion, it is useful to have a list showing different possible interaction terms: see Table 7.1.

Chapter 8

Spontaneous symmetry breaking

The theory of electromagnetism exhibits a gauge invariance, or more elaborately, a local gauge invariance. This can be maintained even if interactions are introduced. In fact, quantum electrodynamics, which is the first quantum field theory to have been fully developed, involves fermions coupled to the electromagnetic field in a gauge invariant way. The quantization procedure is complicated because of the gauge invariance and needs gauge fixing, as we have seen. The gauge field is massless and has only two transverse degrees of freedom as in the free theory. However, gauge interactions can drastically alter the theory. If a scalar field is coupled to the gauge field in a gauge invariant way, normally the situation will be similar, but special forms of the potential may cause the global gauge invariance to be broken. This happens when the scalar field acquires a nonvanishing vacuum expectation value in contrast to the normal situation where the vacuum expectation value of a field vanishes. A nonvanishing vacuum expectation value corresponds to one of an infinite set which is invariant under global gauge transformations. The choice of a single element of the set breaks the invariance. This kind of symmetry breaking is called spontaneous symmetry breaking. This is similar to what happens in ferromagnetism. The elementary magnets are randomly oriented and result in zero total magnetic moment, but on cooling, the elementary magnets become aligned in some direction, producing a total magnetic moment. The selection of a direction breaks the rotational symmetry.

8.1   Abelian Higgs model

A charged scalar field may be described by the Lagrangian density

$$\mathcal{L} = \partial_\mu \phi^* \partial^\mu \phi - m^2 \phi^* \phi - \lambda (\phi^* \phi)^2, \qquad (8.1)$$

where ϕ stands for a complex scalar field

$$\phi = \frac{1}{\sqrt{2}}(\phi_1 + i\phi_2), \tag{8.2}$$

with ϕ_1, ϕ_2 normalized in the usual way. This theory has an internal symmetry. The Lagrangian density is invariant under the global transformation

$$\phi \rightarrow e^{-i\theta}\phi \tag{8.3}$$

and this leads to a conserved current

$$J_\mu = \frac{\partial_\mu \phi^* \phi - \phi^* \partial_\mu \phi}{i} \tag{8.4}$$

and a conserved charge

$$Q = \int d^3\vec{x} \frac{\partial_0 \phi^* \phi - \phi^* \partial_0 \phi}{i} = \int d^3\vec{x} \frac{\pi\phi - \pi^*\phi^*}{i} \tag{8.5}$$

which upon quantization satisfies

$$[\phi, Q] = \phi. \tag{8.6}$$

The potential

$$m^2 \phi^* \phi + \lambda(\phi^* \phi)^2 \tag{8.7}$$

has a minimum at $\phi = 0$, which is invariant under the above symmetry transformation. This is shown in Figure 8.1, where the potential is plotted against ϕ, which for simplicity is represented as a real variable.

Suppose now that instead of real m, one has an imaginary m. This corresponds to a potential

$$-\mu^2 \phi^* \phi + \lambda(\phi^* \phi)^2 \tag{8.8}$$

with real μ. The minimum of this potential is at values of ϕ with

$$\phi^* \phi = \frac{\mu^2}{2\lambda}, \tag{8.9}$$

so that the solution for ϕ is complex in general,

$$\phi = \frac{\mu}{\sqrt{2\lambda}} e^{i\alpha}, \tag{8.10}$$

with a phase α which changes under the above transformation. This is shown in Figure 8.2, where the potential is plotted against ϕ, which for simplicity is represented as a real variable.

The parameter can take a continuum of values, so that the minimum is not unique: there is an infinity of them. The choice of any one minimum breaks

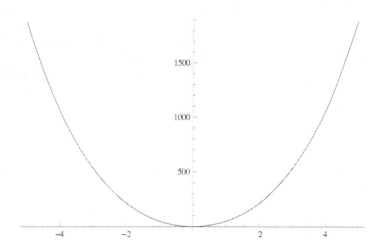

FIGURE 8.1: Symmetric potential.

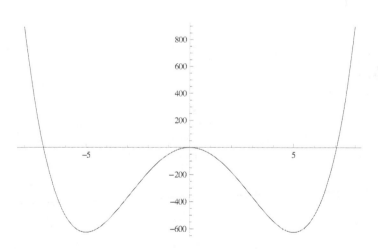

FIGURE 8.2: Symmetric potential with multiple minima.

the symmetry. One may say that in this case of real μ, the ground state or vacuum state of the theory breaks the symmetry. Note that this occurs for real μ only, and not for the more common potential with a real mass term, where the minimum is uniquely at $\phi = 0$. As the Lagrangian density is symmetric and only the ground state breaks it, the symmetry breaking is to be distinguished from explicit symmetry breaking in the action and is referred to as spontaneous symmetry breaking, like spontaneous magnetization.

The field ϕ will have excitations and may be expanded about the minimum:

$$\phi = (\frac{\mu}{\sqrt{2\lambda}} + \rho)e^{i\alpha}. \tag{8.11}$$

Instead of a complex field ϕ or its real and imaginary parts, one may use the two real fields ρ, α corresponding to the modulus and the phase. It is interesting to see that the kinetic part

$$\partial_\mu \phi^* \partial^\mu \phi = \partial_\mu \rho \partial^\mu \rho + (\frac{\mu}{\sqrt{2\lambda}} + \rho)^2 \partial_\mu \alpha \partial^\mu \alpha \tag{8.12}$$

shows up the two real fields, though with an unusual normalization, and interactions between the two fields are already visible here. The potential can also be expressed in terms of these fields:

$$\lambda(\rho^2 + \mu\sqrt{\frac{2}{\lambda}}\rho)^2 - \frac{\mu^4}{4\lambda}, \tag{8.13}$$

which includes a mass term for ρ but nothing for α. Thus one appears to have a scalar field with mass $\sqrt{2}\mu$ and a massless scalar field. A massless scalar field appearing in a theory with spontaneous breaking of a continuous symmetry is called a Goldstone boson, and α, which is directly involved in the symmetry transformation, appears to be just that. The necessary appearance of such a massless scalar is referred to as Goldstone's theorem. The masslessness is related to the fact that there is no change in the potential when one moves between the different solutions corresponding to the minimum.

However, the situation is different in the presence of a gauge interaction. A gauge invariant interaction is obtained by minimally coupling a gauge field, i.e., by replacing the above Lagrangian density by

$$\begin{aligned}\mathcal{L} &= (\partial_\mu + ieA_\mu)\phi^*(\partial^\mu - ieA^\mu)\phi + \mu^2\phi^*\phi - \lambda(\phi^*\phi)^2 \\ &- \frac{1}{4}(\partial_\mu A_\nu - \partial_\nu A_\mu)(\partial^\mu A^\nu - \partial^\nu A^\mu).\end{aligned} \tag{8.14}$$

This Lagrangian density is invariant under the local gauge transformation

$$\phi \to e^{-i\theta}\phi, \quad A_\mu \to A_\mu - \frac{1}{e}\partial_\mu\theta, \tag{8.15}$$

where θ can now be a function of spacetime coordinates.

The phase α can be changed by a gauge transformation, which implies

$$\alpha \to \alpha - \theta. \tag{8.16}$$

This means that this field can be removed by a gauge transformation

$$\phi \to e^{-i\alpha}\phi, \quad A_\mu \to A_\mu - \frac{1}{e}\partial_\mu\alpha, \tag{8.17}$$

and is not a physical field. But a degree of freedom cannot simply vanish. It must show up in a different way. Let us look at the e^2 term in the Lagrangian density. It reads

$$e^2\phi^*\phi A_\mu A^\mu = e^2(\frac{\mu^2}{2\lambda} + 2\frac{\mu}{\sqrt{2\lambda}}\rho + \rho^2)A_\mu A^\mu. \tag{8.18}$$

Apart from interaction terms, this contains a mass term

$$\frac{e^2\mu^2}{2\lambda}A_\mu A^\mu \tag{8.19}$$

for the gauge field which the free theory for that field does not have. Hence something has happened to the gauge field: it has acquired a mass! This means that the Gauss law operator contains a term involving A_0, so that the primary constraint

$$\pi_0 = 0 \tag{8.20}$$

and the corresponding secondary constraint, the Gauss law operator

$$\nabla.\vec{E} + e^2 A_0(\frac{\mu}{\sqrt{2\lambda}} + \rho)^2, \tag{8.21}$$

have a nonvanishing Poisson bracket and are second class. This means that gauge fixing conditions cannot be picked: there is no gauge invariance left. The number of constraints is reduced from the usual four – including conjugates for the two first class constraints – to two. Hence there is an extra degree of freedom in the massive vector field. While the neutral massive boson ρ survives, the phase α metamorphoses to the longitudinal part of the massive vector boson. It is often said that the gauge particle becomes massive by swallowing the would-be Goldstone boson! This process is called the Higgs mechanism. The surviving massive scalar ρ is called the Higgs boson.

The model we have considered above shows A_μ to acquire a mass because for simplicity we have coupled the gauge field to the charged scalar field with a symmetry breaking potential. This mechanism is believed to provide a mass to the weak gauge bosons in the real world, where the electromagnetic field does not acquire any mass but the nonabelian electroweak gauge fields, which are coupled to the scalar field, acquire mass by a version of the mechanism.

8.2   Nonabelian Higgs model

To understand the nonabelian case, we consider a complex scalar field which is a doublet under an $SU(2)$ group and couple it to nonabelian gauge fields corresponding to this $SU(2)$ and the overall phase $U(1)$, forming a $U(2)$. The Lagrangian density for the set of scalar fields may be written as

$$\mathcal{L} = \partial_\mu \phi^\dagger \partial^\mu \phi + \mu^2 \phi^\dagger \phi - \lambda (\phi^\dagger \phi)^2. \qquad (8.22)$$

The minimum of the potential occurs when

$$\phi^\dagger \phi = \frac{\mu^2}{2\lambda}, \qquad (8.23)$$

so that ϕ is of the form

$$\phi = V \begin{pmatrix} \frac{\mu}{\sqrt{2\lambda}} \\ 0 \end{pmatrix}, \qquad (8.24)$$

where V is a unitary 2×2 matrix. All ϕ of this form are at the minimum of the potential, but the theory has to select a specific ϕ. This respects neither the $SU(2)$ nor the $U(1)$ symmetry. The matrix V has four arbitrary parameters, which may be thought of as three $SU(2)$ parameters and one $U(1)$ parameter. However, there is a one parameter combination of rotations by diagonal $SU(2)$ matrices and $U(1)$ which leave the above column vector invariant because they affect only the vanishing lower component and not the nonvanishing upper component. Hence there are only three parameters in the above ground state. More explicitly, with $A = 1, 2$,

$$\phi_A = V_{A1} \frac{\mu}{\sqrt{2\lambda}} \qquad (8.25)$$

and the four real variables in V_{A1} satisfy the normalization condition

$$V_{A1}^* V_{A1} = 1. \qquad (8.26)$$

The field ϕ may be expanded about a minimum in the form

$$\phi = V \begin{pmatrix} \frac{\mu}{\sqrt{2\lambda}} + \rho \\ 0 \end{pmatrix}, \qquad (8.27)$$

with the three parameters mentioned above and ρ allowed to be dependent on spacetime coordinates. The kinetic term in the Lagrangian density becomes

$$\partial_\mu \phi^\dagger \partial^\mu \phi = \partial_\mu \rho \partial^\mu \rho + (\frac{\mu}{\sqrt{2\lambda}} + \rho)^2 \partial_\mu V_{A1}^* \partial^\mu V_{A1}. \qquad (8.28)$$

This has four fields as expected: ρ and the three independent elements of the

V_{A1}. Thus the representation of the complex doublet ϕ of fields in terms of ρ and V does not corrupt the field content.

The ϕ part of the Lagrangian density in the presence of the gauge interactions becomes

$$\mathcal{L} = [(\partial_\mu - iA_\mu)\phi]^\dagger (\partial^\mu - iA^\mu)\phi + \mu^2 \phi^\dagger \phi - \lambda(\phi^\dagger \phi)^2. \qquad (8.29)$$

Here A_μ is a hermitian 2×2 matrix incorporating the $SU(2)$ and $U(1)$ gauge fields. There are two separate coupling constants for $SU(2)$ and $U(1)$ which are absorbed in A_μ for convenience:

$$A_\mu = gA_\mu^a \sigma^a/2 + g'A_\mu', \qquad (8.30)$$

with σ^a denoting the three Pauli matrices. The Lagrangian density is invariant under gauge transformations

$$\phi \to U\phi, A_\mu \to UA_\mu U^{-1} - i(\partial_\mu U)U^{-1}. \qquad (8.31)$$

In terms of V, ρ, the transformation is

$$V_{A1} \to U_{AB}V_{B1}, \quad \rho \to \rho. \qquad (8.32)$$

Therefore, V may be gauged away, i.e., gauge transformed to the identity matrix. However, its three degrees of freedom show up elsewhere. The Lagrangian density has a term

$$\phi^\dagger A^\mu A_\mu \phi = (A^\mu A_\mu)_{11}(\frac{\mu}{\sqrt{2\lambda}} + \rho)^2, \qquad (8.33)$$

which includes a mass term $\frac{\mu^2}{2\lambda}(A^\mu A_\mu)_{11}$ for the gauge fields. As we have seen before, this makes gauge fields massive. But there are four gauge fields present, necessitating care. We see that

$$(A^\mu A_\mu)_{11} = \frac{g^2}{4}A^{a\mu}\Lambda_\mu^a + gg'A_\mu^3 A'^\mu + g'^2 A'^\mu A_\mu', \qquad (8.34)$$

which shows that the $SU(2)$ and $U(1)$ gauge fields become mixed. This quadratic combination of the gauge fields can be diagonalized, yielding eigenvalues $\frac{g^2}{4}, \frac{g^2}{4}, \frac{g^2}{4} + g'^2, 0$. This means that two gauge bosons acquire equal mass, one acquires a higher mass and one remains massless. Since three degrees of freedom from V are removed, it is reasonable that three of the four gauge bosons become massive by the Higgs mechanism. These could be weak interaction mediators. One gauge field remains massless: it is a combination of A_μ^3 and A_μ'. This corresponds to the electromagnetic field. One may say that the electromagnetic gauge invariance is not spontaneously broken. The ground state is invariant under a one parameter combination of $SU(2)$ and $U(1)$ transformations, as mentioned earlier. The $SU(2) \times U(1)$ gauge symmetry is broken spontaneously to a $U(1)$, but it is different from the original

$U(1)$ and is the combination of the original $U(1)$ with a part of the $SU(2)$ which corresponds to the massless gauge boson.

From the four scalar field degrees of freedom, only ρ, which is gauge invariant, survives. Noting that the potential term continues to contain a mass term for ρ and in fact remains unchanged when we pass from the $U(1)$ to this nonabelian case, we conclude that once again this is a massive boson, the Higgs boson.

Chapter 9

Breaking of chiral symmetry on quantization: The anomaly

The chiral symmetry discussed earlier is visible only at the classical level and is not respected by regularizations. Loop integrals arising in the perturbative treatment of quantum field theory need to be regularized before sense can be made of them. There are several ways of carrying out regularizations. Simple procedures like large momentum cutoff are not gauge invariant, so that more elaborate procedures have to be followed. However, there is no regularization that respects both chiral invariance and the phase or global gauge invariance. At least a part of the symmetry shown by the classical theory is broken in the quantum theory. This phenomenon is called an anomaly. In the case of chiral symmetry, which is broken upon quantization of the field theory, one refers to a chiral anomaly. This anomaly may be calculated by using a regularization. We shall study three approaches to the anomaly.

9.1 Pauli–Villars regularization

The Pauli–Villars procedure respects gauge invariance. Here the physical field is supplemented by a regulator field which is made infinitely massive at the end of calculations. This method is suitable for low order calculations, but higher order calculations with complicated diagrams require a generalized version with several regulator fields. Some relations have to be satisfied between the masses of the regulator fields and unusual statistics have to be imposed on them for the infinities to be taken care of.

The Pauli–Villars regularization of a Dirac fermion which is gauged involves the introduction of an extra species of spinors, but with Bose statistics:

$$\mathcal{L}_{\psi,\ PV} = \bar{\psi}(i\gamma^\mu D_\mu - m)\psi + \bar{\phi}(i\gamma^\mu D_\mu - M)\phi. \tag{9.1}$$

Here, ϕ stands for a hypothetical Bose spinor field with mass M. This field or its associated particle is not meant to be seen in experiments. It is a pure artifact, introduced for convenience. Calculations are done with the assumption that this field exists, but at the end of calculations, the mass M is taken to infinity, so that the particle can be taken to decouple from the theory. Only the limit is physically relevant. The physical fermion is taken to have both chiralities and a mass m. Of course this mass has to vanish for chiral symmetry to be present at the classical level, but the calculations are interesting for any mass.

The need for regularization arises because of divergent integrals in perturbative calculations. If one considers a fermion loop with gauge boson lines attached at different points, there will be an integration

$$\int \frac{d^4k}{(2\pi)^4} \text{Tr}[\frac{1}{\slashed{k} - m}\gamma^{\mu_1}\frac{1}{\slashed{k} + \slashed{p}_1 - m}\gamma^{\mu_2}...]. \tag{9.2}$$

A triangle diagram is expected to produce a linear divergence because of the four powers of k in the numerator and three powers coming from the three factors in the denominator. A loop with two gauge boson lines will have only two factors in the denominator and will naïvely be expected to be quadratically divergent in k, but there may be cancellations.

If the theory is regularized in the manner indicated above, each loop will be accompanied by another loop with the physical fermion replaced by the hypothetical particle arising from ϕ. The two diagrams may be combined together and the fermion propagator simply modified by the propagator of the hypothetical particle.

$$\frac{1}{\slashed{k} - m} \to \frac{1}{\slashed{k} - m} - \frac{1}{\slashed{k} - M}. \tag{9.3}$$

For large k, this behaves like

$$\frac{\slashed{k}(m^2 - M^2)}{k^4}, \tag{9.4}$$

so that the divergences of the diagrams are removed at the intermediate calculational stage. The limit $M \to \infty$ has to be taken at the end of all calculations.

In generalized[1] Pauli–Villars regularization, the Lagrangian density is aug-

---

[1]L. D. Faddeev and A. A. Slavnov, *Gauge Fields*, Benjamin-Cummings, Reading (1980).

mented at an intermediate stage to include extra species:

$$\mathcal{L}_{\psi,\,reg} = \bar{\psi}(i\gamma^\mu D_\mu - m)\psi + \sum_j \sum_{k=1}^{|c_j|} \bar{\phi}_{jk}(i\gamma^\mu D_\mu - M_j)\phi_{jk}. \tag{9.5}$$

Here, ϕ_{jk} are regulator spinor fields with fermionic or bosonic statistics, with the integers c_j correspondingly positive or negative and with masses M_j which have to be taken to infinity at the end of any calculations. $|c_j|$ is the number of species with mass M_j.

If one considers a fermion loop with $2n$ gauge boson lines attached at different points, the rationalization of the propagators and calculation of the traces lead to an expression of the form

$$\int \frac{d^4k}{(2\pi)^4} \frac{A_n(k) + m^2 A_{n-1}(k) + ...}{B_{2n}(k) + m^2 B_{2n-1}(k) + ...}, \tag{9.6}$$

where the dependence on other momenta has been suppressed. This leads to

$$\int \frac{d^4k}{(2\pi)^4} \Big[\frac{A_n(k)}{B_{2n}(k)} + m^2\Big(\frac{A_{n-1}(k)}{B_{2n}(k)} - \frac{A_n(k)B_{2n-1}(k)}{B_{2n}^2(k)}\Big) + ...\Big]. \tag{9.7}$$

If the regulator spinors labeled by j are also taken into account, one gets a sum over j. The term corresponding to j has a multiplicity factor $|c_j|$ and a statistics factor which converts this to c_j. If one sets $c_0 = 1$, $M_0 = m$, the original spinor can be labeled by $j = 0$:

$$\int \frac{d^4k}{(2\pi)^4} \sum_{j\geq0} c_j \Big[\frac{A_n(k)}{B_{2n}(k)} + M_j^2\Big(\frac{A_{n-1}(k)}{B_{2n}(k)} - \frac{A_n(k)B_{2n-1}(k)}{B_{2n}^2(k)}\Big) + ...\Big]. \tag{9.8}$$

Note that $A_\ell(k), B_\ell(k)$ behave like $k^{2\ell}$ for large k. The first two terms in the expansion in powers of the masses can be removed by setting

$$\sum_{j\geq0} c_j = 1 + \sum_j c_j = 0, \quad \sum_{j\geq0} c_j M_j^2 = m^2 + \sum_j c_j M_j^2 = 0. \tag{9.9}$$

The subsequent terms have enough powers of k in the denominator to render the integral convergent for $n \geq 1$. This is how this regularization works.

The axial current $\bar{\psi}\gamma^\mu\gamma_5\psi$ corresponding to the global $U(1)$ chiral transformation is generalized in the presence of the extra fields to $\bar{\psi}\gamma^\mu\gamma_5\psi + \sum_{jk} \bar{\phi}_{jk}\gamma^\mu\gamma_5\phi_{jk}$ so that loops involve the regulator spinors too. This regularized current is conserved up to mass terms:

$$\partial^\mu\langle\bar{\psi}\gamma_\mu\gamma_5\psi + \sum_{jk}\bar{\phi}_{jk}\gamma_\mu\gamma_5\phi_{jk}\rangle = 2i\langle\bar{\psi}\gamma_5 m\psi$$

$$+ \sum_{jk}\bar{\phi}_{jk}\gamma_5 M_j\phi_{jk}\rangle. \tag{9.10}$$

The contribution of the regulator fields is

$$2i \sum_j c_j \text{tr} M_j \gamma_5 (i\not{D} - M_j)^{-1}$$
$$= -2i \sum_j c_j \text{tr} M_j^2 \gamma_5 (\not{D}^2 + M_j^2)^{-1}. \tag{9.11}$$

Now

$$\not{D}^2 + M_j^2 = M_j^2 + D^2 + \frac{1}{2} g \sigma^{\mu\nu} F_{\mu\nu}. \tag{9.12}$$

The inverse of this operator can be expanded in powers of F. One can use a representation of the form

$$(A+B)^{-1} = A^{-1} - A^{-1}BA^{-1} + A^{-1}BA^{-1}BA^{-1} - ..., \tag{9.13}$$

with F appearing in B here. Within the trace with γ_5, terms which yield zero trace can be dropped. Four γ matrices are needed, and they have to come from two σ-s. Thus the third term is relevant. In principle, terms with four σ-s etc. could also appear, but they vanish in the limit $M_j \to \infty$ because of high powers of the mass in the denominators. Hence, in the limit, $(M_j^2 + D^2 + \frac{1}{2}g\sigma \cdot F)^{-1}$ can be written here as

$$(M_j^2 + D^2)^{-1} \frac{1}{2} g\sigma \cdot F (M_j^2 + D^2)^{-1} \frac{1}{2} g\sigma \cdot F (M_j^2 + D^2)^{-1}. \tag{9.14}$$

The trace acts in the space of eigenfunctions of $\gamma^\mu D_\mu$, and hence involves summations over spinor and internal indices apart from being a functional trace. The tracing over spinor and internal indices leads to

$$\text{tr} \gamma^5 \sigma^{\mu\nu} \sigma^{\rho\tau} F_{\mu\nu} F_{\rho\tau} = -4\epsilon^{\mu\nu\rho\tau} \text{tr} F_{\mu\nu} F_{\rho\tau} \tag{9.15}$$

in euclidean spacetime. Now the functional trace has to be evaluated. As we are interested in large M_j, the gauge field part of D can be neglected and one can take

$$D_\mu \sim \partial_\mu. \tag{9.16}$$

The eigenfunctions are plane waves and the functional trace becomes

$$\int \frac{d^4p}{(2\pi)^4} \frac{1}{(M_j^2 - p^2)^3}. \tag{9.17}$$

In euclidean spacetime, this integral is nonsingular (unlike the Minkowski case) and has the value

$$\frac{1}{32\pi^2 M_j^2}. \tag{9.18}$$

When use is made of the relation $\sum_j c_j = -1$, all factors combine to produce

$$-\frac{i}{16\pi^2} g^2 \text{tr} \epsilon^{\mu\nu\rho\tau} F_{\mu\nu} F_{\rho\tau} \tag{9.19}$$

in euclidean spacetime. This yields an expression for the divergence of the axial current of the *physical fermion* because the heavy unphysical species decouple in the limit $M_j \to \infty$. The factor of i is dropped when reverting $F_{0\nu}$ to Minkowski spacetime:

$$\partial^\mu \langle \bar{\psi} \gamma_\mu \gamma_5 \psi \rangle = 2i \langle \bar{\psi} \gamma_5 m \psi \rangle - \frac{1}{16\pi^2} g^2 \mathrm{tr} \epsilon^{\mu\nu\rho\tau} F_{\mu\nu} F_{\rho\tau}. \tag{9.20}$$

The second term on the right is not expected from the Lagrangian density. The chiral invariance of the Lagrangian density in the absence of the mass term for the fermion suggests that the right side should be zero apart from corrections involving the mass which are contained in the first term. The extra term is an *anomaly*. It exists also for $m = 0$, when the Lagrangian density has a chiral symmetry. The axial current is classically conserved in that situation. The anomaly signifies a breakdown of this classical chiral symmetry upon quantization due to short distance singularities present in quantum field theory.

It is of interest to see how the anomaly changes if we consider spacetime to be two dimensional. In this case, the expansion stops at the first power of $-\frac{1}{2} g \sigma \cdot F$. The trace is

$$\mathrm{tr} \gamma^5 \sigma^{\mu\nu} F_{\mu\nu} = -\epsilon^{\mu\nu} \mathrm{tr} F_{\mu\nu}. \tag{9.21}$$

The functional trace becomes

$$\int \frac{d^2 p}{(2\pi)^2} \frac{1}{(M_j^2 - p^2)^2} = \frac{1}{4\pi M^2} \tag{9.22}$$

in euclidean spacetime. Hence the final result is

$$\frac{ig}{4\pi} \epsilon^{\mu\nu} \mathrm{tr} F_{\mu\nu}. \tag{9.23}$$

The factor i again is supposed to go away in the passage to (1+1) dimensions.

The anomaly calculation can be adapted to the case of the $SU(n)$ current in the case when the fermion forms an n-tuplet:

$$\partial^\mu \langle \bar{\psi} T^a \gamma_\mu \gamma_5 \psi \rangle + g f^{abc} A^{b\mu} \langle \bar{\psi} T^c \gamma_\mu \gamma_5 \psi \rangle - 2i \langle \bar{\psi} T^a \gamma_5 m \psi \rangle$$
$$= -\frac{1}{16\pi^2} g^2 \mathrm{tr} \epsilon^{\mu\nu\rho\tau} T^a F_{\mu\nu} F_{\rho\tau}. \tag{9.24}$$

This anomaly involves a group theoretic factor $\mathrm{tr}(T^a \{T^b, T^c\})$, which is proportional to what is called d^{abc}. It is nonvanishing for the singlet of $U(1)$ but vanishes for the doublet of $SU(2)$, though not for the triplet of $SU(3)$.

9.1.1 Chiral phases in mass terms

It is possible to introduce phases containing γ^5 in all the mass terms. Suppose the physical mass m is replaced by $m e^{i\theta \gamma_5}$ and the regulator masses

M_j by $M_j e^{i\theta_j \gamma_5}$:

$$\mathcal{L}_{\psi, \, reg}^{[\theta_j]} = \bar{\psi}(i\gamma^\mu D_\mu - m e^{i\theta \gamma_5})\psi + \sum_j \sum_{k=1}^{|c_j|} \bar{\phi}_{jk}(i\gamma^\mu D_\mu - M_j e^{i\theta_j \gamma_5})\phi_{jk}. \tag{9.25}$$

A typical denominator in (9.2) will look like

$$\frac{1}{\not{k} + \not{p} + ... - M_j e^{i\theta_j \gamma_5}} = e^{-i\theta_j \gamma_5/2} \frac{1}{\not{k} + \not{p} + ... - M_j} e^{-i\theta_j \gamma_5/2}. \tag{9.26}$$

Thus each γ^μ in the trace will be sandwiched between two factors of $e^{-i\theta_j \gamma_5/2}$. This applies also to the last γ^μ when the first $e^{-i\theta_j \gamma_5/2}$ is taken to the end by utilizing the cyclic property of the trace. However,

$$e^{-i\theta_j \gamma_5/2} \gamma^\mu e^{-i\theta_j \gamma_5/2} = \gamma^\mu, \tag{9.27}$$

so that all the factors of $e^{-i\theta_j \gamma_5/2}$ cancel out. The conditions (9.9) are unaffected by the phases θ_j which may be present in the original fermion mass term and the regulators.

In the presence of the phases, the divergence equation reads

$$\partial^\mu \langle \bar{\psi}\gamma_\mu \gamma_5 \psi + \sum_{jk} \bar{\phi}_{jk} \gamma_\mu \gamma_5 \phi_{jk} \rangle = 2i \langle \bar{\psi}\gamma_5 m \exp(i\theta\gamma^5)\psi$$

$$+ \sum_{jk} \bar{\phi}_{jk} \gamma_5 M_j \exp(i\theta_j \gamma^5)\phi_{jk} \rangle. \tag{9.28}$$

The contribution of the regulator fields is

$$2i \sum_j c_j \text{tr} M_j e^{i\theta_j \gamma_5} \gamma_5 (i\not{D} - M_j e^{i\theta_j \gamma_5})^{-1}$$
$$= -2i \sum_j c_j \text{tr} M_j^2 \gamma_5 (\not{D}^2 + M_j^2)^{-1}. \tag{9.29}$$

The anomaly is therefore independent of the phases θ_j:

$$\partial^\mu \langle \bar{\psi}\gamma_\mu \gamma_5 \psi \rangle = 2i \langle \bar{\psi}\gamma_5 m \exp(i\theta\gamma^5)\psi \rangle - \frac{1}{16\pi^2} g^2 \text{tr}\epsilon^{\mu\nu\rho\sigma} F_{\mu\nu} F_{\rho\sigma}. \tag{9.30}$$

As the regularized current involves ϕ, chiral transformations corresponding to this current act on ϕ too in addition to ψ:

$$\psi \rightarrow e^{i\alpha\gamma_5}\psi, \quad \bar{\psi} \rightarrow \bar{\psi} e^{i\alpha\gamma_5},$$
$$\phi \rightarrow e^{i\alpha\gamma_5}\phi, \quad \bar{\phi} \rightarrow \bar{\phi} e^{i\alpha\gamma_5}. \tag{9.31}$$

A common phase θ may be introduced in all mass terms if one carries out such a chiral transformation:

$$\mathcal{L}^{[\theta]}_{\psi,\,reg} = \bar{\psi}(i\gamma^\mu D_\mu - me^{i\theta\gamma_5})\psi + \sum_j \sum_{k=1}^{|c_j|} \bar{\phi}_{jk}(i\gamma^\mu D_\mu - M_j e^{i\theta\gamma_5})\phi_{jk}.$$

$$(9.32)$$

This regularized action appears to break parity, but it is symmetric under a chirally rotated parity applied to ψ, ϕ:

$$\psi(\vec{x}) \to \gamma^0 e^{i\theta\gamma_5}\psi(-\vec{x}), \quad \bar{\psi}(\vec{x}) \to \bar{\psi}(-\vec{x})e^{i\theta\gamma_5}\gamma^0$$
$$\phi(\vec{x}) \to \gamma^0 e^{i\theta\gamma_5}\phi(-\vec{x}), \quad \bar{\phi}(\vec{x}) \to \bar{\phi}(-\vec{x})e^{i\theta\gamma_5}\gamma^0.$$

$$(9.33)$$

It also appears to break time reversal invariance. But it is invariant under a chirally rotated time reversal acting on ψ, ϕ. In terms of γ matrices with γ^2 imaginary and the others real,

$$\psi(x^0) \to i\exp(i\theta\gamma_5)\gamma^1\gamma^3\psi(-x^0), \quad \bar{\psi}(x^0) \to \bar{\psi}(-x^0)i\exp(i\theta\gamma_5)\gamma^1\gamma^3$$
$$\phi(x^0) \to i\exp(i\theta\gamma_5)\gamma^1\gamma^3\phi(-x^0), \quad \bar{\phi}(x^0) \to \bar{\phi}(-x^0)i\exp(i\theta\gamma_5)\gamma^1\gamma^3. \quad (9.34)$$

It was pointed out earlier in Chapter 7 that the parity and time reversal transformations are redefined when the mass term has a phase. The above discussion signifies that these transformations are not merely classical symmetries: they are also preserved in the regularized theory. In other words, parity and time reversal are not anomalous and are maintained in the quantum theory even if the mass term has a phase. This eliminates[2] what used to be called the strong CP problem. Of course, CP may still be violated by the $F\tilde{F}$ term of the gauge fields of QCD. As long as there is no experimental indication of CP violation in the strong interaction sector, the coefficient of that term can be set equal to zero. If the phase in the mass term were to contribute to CP violation, the coefficient of the $F\tilde{F}$ term of the gauge fields would have to be tuned to cancel the effect. Such a conspiracy does not have to be invoked because the complex mass term preserves the redefined symmetry. Setting a term equal to zero is usually regarded as natural as it enhances the symmetry of the action. On the other hand, a conspiracy of two effects would appear unnatural.

We have used here only one way of having phases in mass terms. As indicated earlier, phases in the mass terms of the regulators may be chosen arbitrarily without destroying the regularization, as in (9.25). For this general choice, the action is seen to be invariant under a rotated parity transformation, the rotations being different for different j:

$$\psi(\vec{x}) \to \gamma^0 e^{i\theta\gamma_5}\psi(-\vec{x}), \quad \bar{\psi}(\vec{x}) \to \bar{\psi}(-\vec{x})e^{i\theta\gamma_5}\gamma^0$$
$$\phi_{jk}(\vec{x}) \to \gamma^0 e^{i\theta_j\gamma_5}\phi_{jk}(-\vec{x}), \quad \bar{\phi}_{jk}(\vec{x}) \to \bar{\phi}_{jk}(-\vec{x})e^{i\theta_j\gamma_5}\gamma^0.$$

$$(9.35)$$

---

[2]H. Banerjee, D. Chatterjee and P. Mitra, *Phys. Letters* **B573**, 109 (2003).

It is also invariant under a rotated time reversal transformation

$$\psi(x^0) \to i \exp(i\theta\gamma_5)\gamma^1\gamma^3\psi(-x^0),$$
$$\bar{\psi}(x^0) \to \bar{\psi}(-x^0)i \exp(i\theta\gamma_5)\gamma^1\gamma^3,$$
$$\phi_{jk}(x^0) \to i \exp(i\theta_j\gamma_5)\gamma^1\gamma^3\phi_{jk}(-x^0),$$
$$\bar{\phi}_{jk}(x^0) \to \bar{\phi}_{jk}(-x^0)i \exp(i\theta_j\gamma_5)\gamma^1\gamma^3. \tag{9.36}$$

Note that the chiral transformations involved here are different from the chiral transformation corresponding to the regularized axial current.

The invariance under parity and time reversal transformations of (9.25) may also be understood in a different way. Note first that the γ matrices may be chosen independently for the fields ψ and ϕ_j. For ψ, one can set

$$\gamma^\mu e^{i\theta\gamma^5} = \tilde{\gamma}^\mu \tag{9.37}$$

and for the regulator fields ϕ_j one may set

$$\gamma^\mu e^{i\theta_j\gamma^5} = \tilde{\gamma}^\mu_{(j)}. \tag{9.38}$$

These new γ matrices satisfy the required relations

$$\{\tilde{\gamma}^\mu, \tilde{\gamma}^\nu\} = \{\tilde{\gamma}^\mu_{(j)}, \tilde{\gamma}^\nu_{(j)}\} = 2g^{\mu\nu}. \tag{9.39}$$

If the Lagrangian density is rewritten in terms of these matrices, the phases θ, θ_j become invisible. The theory cannot depend on such parameters.

9.2   Measure formulation

Anomalies are often studied in the euclidean functional integral framework. The action for massless quarks is chirally invariant at the classical level, so the breaking of this invariance by the anomaly can only happen through the measure of integration. Now the infinite dimensional measure is not easy to control, but it can be handled by suitable expansion techniques, followed by regularization.

The action

$$S = \int \bar{\psi}(i\gamma^\mu D_\mu - m)\psi \tag{9.40}$$

for a Dirac fermion interacting with gauge fields and possibly having a mass is expressed in terms of expansion coefficients of the fermion fields in eigenfunctions ϕ_n of the hermitian operator $\gamma^\mu D_\mu$, which must form a complete set. Thus the independent fields $\psi, \bar{\psi}$ are expanded as

$$\psi = \sum_n a_n\phi_n, \quad \bar{\psi} = \sum_n \bar{a}_n\phi_n^\dagger, \tag{9.41}$$

and the functional integral is written as

$$Z = \int \mathcal{D}A \prod_n \int da_n \prod_n \int d\bar{a}_n e^{-S}, \tag{9.42}$$

with the measure taken to be the product of the expansion coefficients. This is a definition of the measure. It is gauge invariant in the sense that ψ and ϕ_n transform in the same way under a gauge transformation, as do $\bar{\psi}$ and $\phi_n^\dagger$, making each $a_n, \bar{a}_n$ gauge invariant. For a chiral transformation

$$\psi \to e^{i\alpha\gamma^5}\psi, \quad \bar{\psi} \to \bar{\psi}e^{i\alpha\gamma^5}, \tag{9.43}$$

the expansion coefficients change:

$$
\begin{aligned}
a_n &\to \sum_m \int \phi_n^\dagger e^{i\alpha\gamma^5} \phi_m a_m, \\
\bar{a}_n &\to \sum_m \bar{a}_m \int \phi_m^\dagger e^{i\alpha\gamma^5} \phi_n.
\end{aligned}
\tag{9.44}
$$

The measure changes and there is a Jacobian which has to be calculated after a regularization involving the gauge invariant eigenvalues of $\gamma^\mu D_\mu$. We may write the above equations in the form

$$
\begin{aligned}
a_n &\to \sum_m C_{nm} a_m, \\
\bar{a}_n &\to \sum_m \bar{a}_m \bar{C}_{mn},
\end{aligned}
\tag{9.45}
$$

with matrices $C, \bar{C}$. Noting that for a Grassmann variable a,

$$d(ca) = \frac{1}{c} da, \tag{9.46}$$

as required by the normalization of integrals, we see that the determinants have to appear with negative powers and express the new measure as

$$\prod_n (da_n' d\bar{a}_n') = (\det C)^{-1} (\det \bar{C})^{-1} \prod_n (da_n d\bar{a}_n). \tag{9.47}$$

To first order in α, this becomes

$$\prod_n (da_n' d\bar{a}_n') = [1 - 2i \int d^4x \alpha \sum_n \phi_n^\dagger \gamma^5 \phi_n] \prod_n (da_n d\bar{a}_n). \tag{9.48}$$

Thus one has to evaluate the trace of the operator γ^5 in the space of eigenfunctions of the hermitian operator $\gamma^\mu D_\mu$, but one has to regularize the sum. A possibility is to introduce a factor $\exp(-\slashed{D}^2/\Lambda^2)$ with Λ taken to infinity. This is gauge invariant and screens large eigenvalues. Because of the completeness of the eigenfunctions, one can evaluate the trace in a more convenient basis, involving plane waves. Thus the regularized trace may be written as

$$\lim_{\Lambda \to \infty} \int \frac{d^4 k}{(2\pi)^4} e^{-ikx} \mathrm{tr} \gamma^5 \exp(-\slashed{D}^2/\Lambda^2) e^{ikx}. \tag{9.49}$$

Here, with euclidean spacetime and antihermitian γ matrices,

$$(\gamma^\mu D_\mu)^2 = -D^2 + \frac{1}{2} g \sigma^{\mu\nu} F_{\mu\nu}. \tag{9.50}$$

The plane wave function e^{ikx} may be transferred across D_μ, which produces ik_μ (plus derivatives which are suppressed by Λ):

$$\lim_{\Lambda \to \infty} \int \frac{d^4 k}{(2\pi)^4} \exp(-k^2/\Lambda^2) \mathrm{tr} [\gamma^5 \exp(-\frac{g\sigma^{\mu\nu} F_{\mu\nu}}{2\Lambda^2})]. \tag{9.51}$$

The k integration can be done componentwise and yields a factor $(\sqrt{\pi})^4 \Lambda^4/(2\pi)^4$. The exponential in the matrices can be expanded. A nonzero trace arises from the second power:

$$\frac{g^2}{8\Lambda^4} \mathrm{tr}\, F_{\mu\nu} F_{\rho\sigma} (-4\epsilon^{\mu\nu\rho\sigma}). \tag{9.52}$$

The final result for the trace is

$$-\frac{g^2}{32\pi^2} \mathrm{tr}\, F_{\mu\nu} F_{\rho\sigma} \epsilon^{\mu\nu\rho\sigma}. \tag{9.53}$$

The measure therefore changes by a factor

$$1 + \frac{i\alpha g^2}{16\pi^2} \int d^4 x \mathrm{tr}\, F_{\mu\nu} F_{\rho\sigma} \epsilon^{\mu\nu\rho\sigma} \tag{9.54}$$

for small α, and this exponentiates to

$$\exp(\frac{i\alpha g^2}{16\pi^2} \int d^4 x \mathrm{tr}\, F_{\mu\nu} F_{\rho\sigma} \epsilon^{\mu\nu\rho\sigma}). \tag{9.55}$$

This is the effect of the chiral transformation. An $F\tilde{F}$ term is produced and may be interpreted as being a part of the action. This is the way in which the chiral symmetry is broken and this change of the measure is a symptom of the anomaly.

Using (3.4) for the chiral transformation with small α, and including the change of the action resulting from the change in the measure found above, we see that

$$\partial_\mu (\bar{\psi} \gamma^\mu \gamma^5 \psi) = 2im\bar{\psi}\gamma^5\psi - i\frac{g^2}{16\pi^2} \mathrm{tr}\, F_{\mu\nu} F_{\rho\sigma} \epsilon^{\mu\nu\rho\sigma}. \tag{9.56}$$

This is the euclidean form of the anomaly equation, now obtained in the measure approach.

It is of interest to see how this changes in two dimensional spacetime. The k integration yields $(\sqrt{\pi})^2 \Lambda^2 / (2\pi)^2$. A nonzero trace arises from the first power in the expansion:

$$-\frac{g}{2\Lambda^2} \text{tr } F_{\mu\nu}(-\epsilon^{\mu\nu}). \tag{9.57}$$

The final result for the trace is

$$\frac{g}{8\pi} \text{tr } F_{\mu\nu}\epsilon^{\mu\nu}, \tag{9.58}$$

so that the euclidean anomaly in this spacetime is

$$\frac{ig}{4\pi}\epsilon^{\mu\nu}\text{tr } F_{\mu\nu}.$$

As indicated in the previous section, the anomaly can only be detected if the theory is regularized. While the Pauli–Villars regularization in a generalized form was used in that section, a regularization is used here in the evaluation of the trace of γ^5 in the space of eigenfunctions of $\not{D}$, which naïvely would appear to be zero because the trace of the γ^5 matrix vanishes. This regularization is used here within a calculation which is initially formulated without any explicit regularization. But the measure approach can also be applied in a regularized theory. If the generalized Pauli–Villars regularization is used, the measure will become more complicated as all fields – both physical and unphysical – have to be integrated over. The change of the measure will acquire contributions from all these fields. Now the contribution depends on the statistics of the fields: the determinants come with negative powers for Grassmann variables and positive powers for normal variables associated with bosonic fields. The contribution from the j regulators has a multiplicity factor $|c_j|$ and a statistics factor given by the relative sign of c_j, so that the contribution is proportional to c_j. Thus the net contribution to the change of the measure under a chiral transformation contains, apart from the trace of γ^5, a factor $1 + \sum_j c_j$, which vanishes by (9.9). The measure factor becomes

$$[1 - 2i \int d^4x \alpha (1 + \sum_j c_j) \sum_n \phi_n^\dagger \gamma^5 \phi_n] = 1. \tag{9.59}$$

Thus in this explicitly regularized theory, the measure is invariant under a chiral transformation[3] and the anomaly has to be derived by other means, as already seen.

---

[3] K. Fujikawa, *Int. J. Mod. Phys.* **A16**, 331 (2001).

9.2.1   Chiral phases in mass terms

For a complex mass term, the measure may change. The action is

$$\int \bar{\psi}(i\gamma^\mu D_\mu - me^{i\theta\gamma^5})\psi. \tag{9.60}$$

We know that this action is invariant under parity and time reversal transformations which are obtained by a chiral rotation from the transformations appropriate to the case of a real mass term. Let us therefore take the expansion of the fields provisionally as

$$\psi = e^{-i\beta\gamma^5/2} \sum_n a_n^\beta \phi_n, \quad \bar{\psi} = \sum_n \bar{a}_n^\beta \phi_n^\dagger e^{-i\beta\gamma^5/2} \tag{9.61}$$

with the phase β to be determined later. Then

$$Z = \int \mathcal{D}A \prod_n \int da_n^\beta \prod_n \int d\bar{a}_n^\beta e^{-S}, \tag{9.62}$$

the measure being gauge invariant but potentially parity violating because of the β factor. This is allowed because of the presence of the potentially parity violating θ.

Now the phases θ, β can be removed by chiral transformations $\exp(-i\theta\gamma^5/2), \exp(i\beta\gamma^5/2)$ at the expense of changes in the coefficient of the $F\tilde{F}$ term as above. The effective parity violation parameter will therefore be

$$\bar{\theta} = \theta - \beta \tag{9.63}$$

if there is no $F\tilde{F}$ term to begin with.

Note that under a chiral transformation,

$$a_n^\beta \rightarrow \sum_m \int \phi_n^\dagger e^{i\alpha\gamma^5} \phi_m a_m^\beta,$$

$$\bar{a}_n^\beta \rightarrow \sum_m \bar{a}_m^\beta \int \phi_m^\dagger e^{i\alpha\gamma^5} \phi_n, \tag{9.64}$$

which is the same as for a real mass term, implying that in spite of the presence of the parameter β, one gets the same change of measure for a chiral transformation and hence the same anomaly.

To fix β, we have to understand how the measure should change under parity. Under a *parity operation for gauge fields*,

$$\phi_n(x_0, \vec{x}) \rightarrow \gamma^0 \phi_n(x_0, -\vec{x}), \tag{9.65}$$

$$\phi_n^\dagger(x_0, \vec{x}) \rightarrow \phi_n^\dagger(x_0, -\vec{x})\gamma^0, \tag{9.66}$$

so that

$$\phi_n^\dagger(x_0, \vec{x})e^{-i\beta\gamma^5/2} \rightarrow \phi_n^\dagger(x_0, -\vec{x})e^{-i\beta\gamma^5/2}e^{i\beta\gamma^5}\gamma^0,$$

$$e^{-i\beta\gamma^5/2}\phi_n(x_0, \vec{x}) \rightarrow \gamma^0 e^{i\beta\gamma^5}e^{-i\beta\gamma^5/2}\phi_n(x_0, -\vec{x}). \tag{9.67}$$

TABLE 9.1:   Choice of Phase in Measure.

Measure with $\beta = 0$	Measure with $\beta = \theta$
Agrees with standard measure for real mass term	Reduces to standard measure for real mass term
Standard anomaly	Standard anomaly
Measure invariant under usual parity but not under actual symmetry of action	Measure invariant under actual symmetry of action but not under usual parity
$\bar{\theta} = \theta$	$\bar{\theta} = 0$

Compare this with the parity transformation for fermion fields. One has to realize that the naïve parity transformation is altered in the presence of the chiral phase in the mass term, as seen in Chapter 7. For consistency of the expansions, the two sides should transform in the same way. It is thus seen to be necessary to have

$$\beta = \theta, \qquad\qquad (9.68)$$

and therefore

$$\bar{\theta} = \theta - \beta = 0. \qquad\qquad (9.69)$$

Thus the parameter β in the measure has to be fixed to be equal to the parameter θ in the action. The redefined measure simply respects the altered parity. There is no net parity violation in spite of the apparent violations in the action and in the measure. There is a similar obedience of the redefined time reversal symmetry.

We can compare the situation in the presence of β in the expansion with what would have happened if the usual expansion without β had been retained. This is indicated in Table 9.1.

An equivalent way of looking at these symmetries is as follows. One may consider the fermion action with a real mass term and hence invariant under parity P:

$$S[\psi, \bar{\psi}, A] = S[P\psi, \bar{\psi}P, A^P]. \qquad\qquad (9.70)$$

Here A is the gauge field and A^P its parity transformed form. A chiral transformation $\exp(i\theta\gamma^5/2)$ on the fields generates the complex mass term. More generally, let us consider a chiral transformation $\chi = \exp(i\beta\gamma^5/2)$. The action will change and will be invariant not under P, but under a new transformation:

$$
\begin{aligned}
S_\chi[\psi, \bar{\psi}, A] &\equiv S[\chi\psi, \bar{\psi}\chi, A] \\
&= S[P\chi\psi, \bar{\psi}\chi P, A^P] \\
&= S[\chi\chi^{-1}P\chi\psi, \bar{\psi}\chi P\chi^{-1}\chi, A^P] \\
&= S_\chi[\chi^{-1}P\chi\psi, \bar{\psi}\chi P\chi^{-1}, A^P].
\end{aligned}
\tag{9.71}
$$

This symmetry simply involves P rotated by χ.

What about the fermion regularization or measure? For the action S, the conventional fermion measure, with implicit regularization, is invariant under P:

$$
d\mu[\psi, \bar{\psi}, A] = d\mu[P\psi, \bar{\psi}P, A^P].
\tag{9.72}
$$

But this $d\mu$ does not in general respect the rotated parity symmetry as there is a chiral anomaly:

$$
d\mu[\psi, \bar{\psi}, A] \neq d\mu[\chi^{-1}P\chi\psi, \bar{\psi}\chi P\chi^{-1}, A^P].
\tag{9.73}
$$

But one may manufacture a new measure:

$$
\begin{aligned}
d\mu_\chi[\psi, \bar{\psi}, A] &\equiv d\mu[\chi\psi, \bar{\psi}\chi, A] \\
&= d\mu[P\chi\psi, \bar{\psi}\chi P, A^P] \\
&= d\mu_\chi[\chi^{-1}P\chi\psi, \bar{\psi}\chi P\chi^{-1}, A^P],
\end{aligned}
\tag{9.74}
$$

which formally has the same symmetry as S_χ. From this construction, the new measure has a parameter β. This is a one parameter class of variations of the measure.

If one wishes to generate the θ phase from a real mass term by the chiral transformation χ, one has to take

$$
\beta = \theta.
\tag{9.75}
$$

The corresponding choice of the measure $d\mu_\chi$ respects the parity symmetry of the complex fermion action. Parity is not broken by the combination of the complex mass term and this measure in the functional integral. Similar arguments apply for time reversal. These may be broken only by a vacuum angle, i.e., an $F\tilde{F}$ term of the action.

These constructions are in the presence of gauge fields: anomalies have been implicitly taken into account as it is only because of anomalies that the measure $d\mu_\chi$ depends on χ.

This θ dependence of the measure is important in understanding the absence of any strong CP problem. If the naïve measure were used and the mass term rid of its phase θ by a chiral transformation, it would then generate an $F\tilde{F}$ term and thus cause parity and time reversal violation. Because of the matching of the phases of the mass term and the θ dependent measure, the contribution to the $F\tilde{F}$ term coming from the complex mass term is canceled by the contribution arising upon the removal of the phase θ in the measure, so that there is no trace of θ in the $F\tilde{F}$ term in the end. It is important that the measure is sought to respect the symmetry of the action. While it cannot be made chirally invariant, it can be made parity and time reversal invariant. That is why there is a chiral anomaly but no parity or time reversal anomaly. The naïve measure respects the naïve parity and time reversal symmetries, i.e., the symmetries occurring for real mass, while the proper measure respects the rotated parity and time reversal symmetries occurring for the appropriate complex mass. We have also seen in (9.59) that if the theory is regularized, say by the Pauli–Villars method, a chiral transformation carried out to remove the phase in the mass term does *not* generate an $F\tilde{F}$ term from the measure. If there still is any parity or time reversal violation in strong interactions, it is due to an intrinsic $F\tilde{F}$ term and does not come from the fermions.

9.3 Zeta function regularization

In this section we employ a new regularization. The determinant of a matrix is the product of its eigenvalues. For an operator, the product of the eigenvalues has to be regularized. The zeta function regularization is occasionally used in quantum field theory. The chiral anomaly in vector gauge theories can also be evaluated by using this regularization.

The zeta function of an operator X is defined for a parameter s,

$$\zeta(s, X) \equiv \mathrm{Tr}(X^{-s}). \tag{9.76}$$

If the eigenvalues are λ, this becomes

$$\zeta(s, X) - \sum(\lambda^{-s}), \tag{9.77}$$

so that

$$\zeta'(s, X) = -\sum(\ln \lambda \lambda^{-s}), \tag{9.78}$$

and

$$\zeta'(0, X) = -\sum(\ln \lambda) = -\ln \prod \lambda = -\ln \det X. \tag{9.79}$$

This is a definition of the determinant. The eigenvalues are assumed to be positive in this definition.

Note that the Dirac operator

$$\mathcal{D} = i\slashed{D} - m \tag{9.80}$$

is neither hermitian nor antihermitian if the mass is nonvanishing. A *positive* operator may be constructed for the zeta function by going over to the Laplacian from the Dirac operator:

$$\Delta = [i\not{D} - m]^\dagger [i\not{D} - m]. \tag{9.81}$$

The operator may be said to have been squared in the process, and a square root has to be included in the definition of the determinant.

Now *antihermitian* γ matrices are used in euclidean spacetime. Then

$$\Delta = (\not{D})^2 + m^2. \tag{9.82}$$

The zeta function of this operator is

$$\zeta(s, \Delta) \equiv \text{Tr}(\Delta^{-s}), \tag{9.83}$$

and the regularized logarithm of the functional integral is defined in the limit of $s \to 0$ for some μ as

$$\ln Z \equiv -\frac{1}{2}\zeta'(0, \Delta) - \frac{1}{2}\ln \mu^2 \zeta(0, \Delta). \tag{9.84}$$

The square root is introduced to compensate for the squaring in the construction of Δ. It is to be noted that the determinant is defined only for the product Δ and not for the linear Dirac operators.

In the important case of a chiral phase in the mass term,

$$\Delta = [i\not{D} - m\exp(i\theta\gamma^5)]^\dagger [i\not{D} - m\exp(i\theta\gamma^5)] = (\not{D})^2 + m^2, \tag{9.85}$$

the same as before. The regularized determinant is unchanged and is independent of θ, depending on the gauge fields only through the operator Δ and is therefore invariant under symmetry transformations of the gauge field A. The phase of the mass term thus cannot cause any CP violation in this regularization.

Let us imagine the determinant of a Dirac operator when $\not{D}$ has only a finite number of zero modes and no other eigenvalue. The anticommutation of $\not{D}$ and γ^5 allows the zero modes to be chosen to have definite chirality. The mass term produces a factor of $\exp(i\theta\gamma^5)$ for each zero mode, leading to a product $\exp(i\theta\nu)$, where ν is the number of positive chirality zero modes *minus* the number of negative chirality zero modes. Here ν depends on the gauge field involved. This factor occurring in the determinant of the Dirac operator is canceled by its conjugate in the determinant of the operator Δ. When the number of modes becomes infinite, a regularization is necessary. The zeta function regularization ensures that the determinant of Δ is θ independent. As we show below, it still reproduces the anomaly, indicating the decoupling of θ from the anomaly.

For calculating Green functions, one has to introduce source terms in the standard way. The determinant may be defined by the method indicated above, while the factor involving the source is separate. This factor explicitly involves θ through the Dirac operator. Thus fermionic Green functions depend on θ for complex mass terms in the zeta function approach. The fermion field is not chirally invariant. But there is no contribution to the θ dependence from fermion loops in the determinant.

If one wishes to derive the anomaly equation in this regularization, one has to add source terms for the fermionic operators occurring in that equation to the Dirac operator. This produces the modified operator

$$[i\slashed{D} - iQ\!\!\!\!/\,\gamma^5 - m - K\gamma^5], \tag{9.86}$$

with $Q^\mu(x)$ coupling to the axial current and K to the pseudoscalar density including the phase. Now in the presence of the phase in the mass term, the parity symmetry transformation of the action is chirally rotated, so that $\bar{\psi}\exp(i\theta\gamma^5)\psi$ is a scalar and $\bar{\psi}\gamma^5\exp(i\theta\gamma^5)\psi$ is a pseudoscalar. Then the pseudoscalar source too will have the chiral phase factor. However, we shall use a real mass term and the modified Laplacian[4]

$$
\begin{aligned}
\Delta' &= [i\slashed{D} - iQ\!\!\!\!/\,\gamma^5 - m - K\gamma^5]^\dagger \\
&\quad [i\slashed{D} - iQ\!\!\!\!/\,\gamma^5 - m - K\gamma^5] \\
&= (\slashed{D})^2 + m^2 + K^2 - (Q\!\!\!\!/\,)^2 - Q\!\!\!\!/\,\gamma^5\slashed{D} - \slashed{D}Q\!\!\!\!/\,\gamma^5 + 2mK\gamma^5 \\
&\quad + i(\slashed{D} - Q\!\!\!\!/\,\gamma^5)K\gamma^5 - iK\gamma^5(\slashed{D} - Q\!\!\!\!/\,\gamma^5).
\end{aligned}
\tag{9.87}
$$

This Δ' operator helps us define a modified Z' and hence the expectation value of the axial current operator

$$\langle\bar{\psi}\gamma_\mu\gamma_5\psi\rangle = i\frac{\delta \ln Z'}{\delta Q^\mu(x)}\Big|_{Q=K=0} = -\frac{i}{2}\frac{\delta\zeta'(0,\Delta')}{\delta Q^\mu(x)}\Big|_{Q=K=0}. \tag{9.88}$$

Now let ϕ_n be eigenfunctions with eigenvalues λ_n for Δ, primed where required (see below):

$$\Delta\phi_n = \lambda_n\phi_n, \quad \Delta'\phi_n' = \lambda_n'\phi_n'. \tag{9.89}$$

By the Hellman–Feynman theorem of perturbation theory,

$$
\begin{aligned}
\frac{\delta\zeta(s,\Delta')}{\delta Q^\mu(x)}\Big|_{Q=K=0} &= \sum_n \frac{\delta\lambda_n'^{-s}}{\delta Q^\mu(x)}\Big|_{Q=K=0} \\
&= -s\sum_n \lambda_n^{-s-1}\frac{\delta\lambda_n'}{\delta Q^\mu(x)}\Big|_{Q=K=0} \\
&= -s\sum_n \lambda_n^{-s-1}\int d^4w\,\phi_n^\dagger(w)\frac{\delta\Delta'}{\delta Q^\mu(x)}\Big|_{Q=K=0}\phi_n(w) \\
&= s\sum_n \lambda_n^{-s-1}\phi_n^\dagger(x)[\gamma_\mu\gamma^5\slashed{D} - \overleftarrow{\slashed{D}}\gamma_\mu\gamma^5]\phi_n(x). \tag{9.90}
\end{aligned}
$$

---

[4]M. Reuter, *Phys. Rev.* **D31**, 1374 (1985).

When taking the divergence one can simplify the expression if one chooses the ϕ_n as eigenfunctions of $\not{D}$ in addition to Δ:

$$\not{D}\phi_n = \alpha_n\phi_n. \tag{9.91}$$

This is possible when

$$\alpha_n^2 + m^2 = \lambda_n \tag{9.92}$$

because $\not{D}$ is hermitian and its square differs by m^2 from Δ. One finds

$$
\begin{aligned}
\partial^\mu\langle\bar\psi\gamma_\mu\gamma_5\psi\rangle &= 2i\sum_n[s\lambda_n^{-s}\phi_n^\dagger(x)\gamma^5\phi_n(x)]'|_{s\to0} \\
&\quad -2im^2\sum_n[s\lambda_n^{-s-1}\phi_n^\dagger(x)\gamma^5\phi_n(x)]'|_{s\to0}.
\end{aligned} \tag{9.93}
$$

Let us consider the first term. There is a factor $\operatorname{tr}\gamma^5\sum_n\lambda_n^{-s}\phi_n(x)\phi_n^\dagger(y)$, with $y\to x$. Here λ_n can be replaced by the operator Δ whose eigenvalue it is, because of the presence of the eigenfunction ϕ_n. The sum over n of $\phi_n\phi_n^\dagger$ can also be written in terms of plane waves, which makes it easier to handle. The fractional power of Δ can be cast in the form of an integral. Thus one has to consider

$$\frac{1}{\Gamma(s)}\int_0^\infty dt\,t^{s-1}\int\frac{d^4k}{(2\pi)^4}e^{-ik\cdot x}e^{-\Delta t}e^{ik\cdot x}. \tag{9.94}$$

Now $\Delta = -D^2 + m^2 + \frac{g}{2}\sigma^{\mu\nu}F_{\mu\nu}$. Each D_μ can be replaced by ik_μ, apart from derivative corrections which do not contribute in the $s\to0$ limit. The k integration in euclidean space yields a factor of $(\sqrt\pi)^4/(2\pi\sqrt t)^4 = 1/(16\pi^2t^2)$. The factor $e^{-\frac{1}{2}g\sigma^{\mu\nu}F_{\mu\nu}t}$ is expanded as the exponential series. Many terms vanish when the trace with γ^5 or the $s\to0$ limit is taken. The nonvanishing term is the one with power t^2. The $1/t^2$ and t^2 cancel and the t integral reduces to m^{-2s}. There was a differentiation with respect to s, which is supposed to be carried out after multiplication by s. This simply yields 1 in the limit $s\to0$. The trace reduces to

$$\frac{g^2}{8}\operatorname{tr}\gamma^5\sigma^{\mu\nu}\sigma^{\rho\tau}F_{\mu\nu}F_{\rho\tau}. \tag{9.95}$$

When all these factors are taken together, one finds

$$-\frac{ig^2}{16\pi^2}\operatorname{tr}\epsilon^{\mu\nu\rho\tau}F_{\mu\nu}F_{\rho\tau}. \tag{9.96}$$

The second term of (9.93) can be rewritten[5] in a familiar form. Note that

$$
\begin{aligned}
\langle\bar\psi\gamma_5\psi\rangle &= -\frac{\delta\ln Z'}{\delta K(x)}\Big|_{Q=K=0} \\
&= \frac{1}{2}\frac{\delta\zeta'(0,\Delta')}{\delta K(x)}\Big|_{Q=K=0}.
\end{aligned} \tag{9.97}
$$

---

[5]P. Mitra, *Europhys. J.* **C72**, 2024 (2012).

Now as in the case of the axial current operator, one has

$$
\begin{aligned}
\frac{\delta \zeta(s, \Delta')}{\delta K(x)}\Big|_{Q=K=0} &= \sum_n \frac{\delta \lambda_n'^{-s}}{\delta K(x)}\Big|_{Q=K=0} \\
&= -s \sum_n \lambda_n^{-s-1} \frac{\delta \lambda_n'}{\delta K(x)}\Big|_{Q=K=0} \\
&= -s \sum_n \lambda_n^{-s-1} \int d^4 w \, \phi_n^\dagger(w) \frac{\delta \Delta'}{\delta K(x)}\Big|_{Q=K=0} \phi_n(w) \\
&= -s \sum_n \lambda_n^{-s-1} \phi_n^\dagger(x) [-i\gamma^5 \not{D} \\
&\qquad\qquad -i \overleftarrow{\not{D}} \gamma^5 + 2m\gamma^5] \phi_n(x).
\end{aligned} \tag{9.98}
$$

This may be simplified by recalling that $\phi_n(x)$ is an eigenfunction of the operator $\not{D}$ in addition to $\Delta = (\not{D})^2 + m^2$, as chosen earlier. One finds

$$
\langle \bar\psi \gamma_5 \psi \rangle = -m \sum_n [s\lambda_n^{-s-1} \phi_n^\dagger(x) \gamma^5 \phi_n(x)]'|_{s\to 0}, \tag{9.99}
$$

so that

$$
\partial^\mu \langle \bar\psi \gamma_\mu \gamma_5 \psi \rangle = 2im \langle \bar\psi \gamma_5 \psi \rangle - \frac{ig^2}{16\pi^2} \mathrm{tr}\, \epsilon^{\mu\nu\rho\sigma} F_{\mu\nu} F_{\rho\sigma}. \tag{9.100}
$$

This is the anomaly equation in euclidean space once again.

It is interesting to see how this changes if we go to a two-dimensional spacetime. The k integration in two dimensional euclidean space yields a factor $(\sqrt{\pi})^2/(2\pi\sqrt{t})^2 = 1/(4\pi t)$. When the exponential in t is expanded, the linear term is the one that matters in this case. The t cancels the $1/t$. The trace leads to

$$
-\frac{g}{2} \mathrm{tr}\, \gamma^5 \sigma^{\mu\nu} F_{\mu\nu} = \frac{g}{2} \epsilon^{\mu\nu} \mathrm{tr}\, F_{\mu\nu}, \tag{9.101}
$$

and together with the factor of $2i$, one obtains

$$
\frac{ig}{4\pi} \epsilon^{\mu\nu} \mathrm{tr}\, F_{\mu\nu} \tag{9.102}
$$

as the anomaly in two dimensional euclidean spacetime. The factor of i disappears on continuation to (1+1) dimensions. The anomaly is thus proportional to the electric field in this spacetime.

9.4   Chiral gauge theory

A chiral version of the anomaly equation can be obtained by combining it with the covariant conservation equation of the vector current. The mass term has to be dropped for the chiral theory, which has only one chirality. The gauge interaction is chiral, i.e., the gauge field interacts with only one chirality, so this is a chiral gauge theory. This is unlike the common gauge theories like electrodynamics, where the gauge field interacts with a vector current involving both chiralities. It is also possible to have gauge fields interacting with axial currents. If we consider only the left chirality to be present,

$$\partial^\mu \langle \bar{\psi}_L T^a \gamma_\mu \psi_L \rangle + g f^{abc} A^{b\mu} \langle \bar{\psi}_L T^c \gamma_\mu \psi_L \rangle = \frac{1}{32\pi^2} g^2 \mathrm{tr} \epsilon^{\mu\nu\rho\tau} T^a F_{\mu\nu} F_{\rho\tau}.$$

(9.103)

This is what is known as the covariant form of the anomaly in the chiral gauge theory. It corresponds to a gauge covariant regularization. An alternative form which is not gauge covariant arises through the construction of an effective action. Functional differentiation of the effective action with respect to the gauge field yields a current, whose covariant divergence defines an anomaly. The two anomalies are related, but not equal. The alternative anomaly is *consistent* with a relation that has to be satisfied if it is the covariant derivative of a current derived from an effective action. The above approach does not involve evaluating an effective action, so the covariant anomaly is not expected to be *consistent* in that sense. However, for the equations of motion of gauge fields to be satisfied, the anomaly in a chiral gauge theory has to vanish. As indicated above, this happens if d^{abc} vanishes. The consistent anomaly too vanishes when the covariant one vanishes, namely, when this condition is satisfied.

To study these issues, we first note that the gauge transformation of a covariantly regularized current should be like

$$\delta J_\mu^a = g f^{abc} \theta^b J_\mu^c.$$

(9.104)

Now gauge transformations are caused by changes of the gauge field, on which the expectation value of the current depends. Thus,

$$
\begin{aligned}
\delta J_\mu^a &= \int d^4 y \frac{\delta J_\mu^a}{\delta A_\nu^b(y)} \delta A_\nu^b(y) \\
&= \int d^4 y \frac{\delta J_\mu^a}{\delta A_\nu^b(y)} [-\partial_\nu \theta^b(y) + g f^{bcd} \theta^c(y) A_\nu^d(y)] \\
&= \int d^4 y [\partial_\nu^y \frac{\delta J_\mu^a}{\delta A_\nu^b(y)} \theta^b(y) + g f^{bdc} A_\nu^d(y) \frac{\delta J_\mu^a}{\delta A_\nu^c(y)} \theta^b(y)] \\
&= \int d^4 y L^b(y) J_\mu^a(x) \theta^b(y),
\end{aligned}
$$

(9.105)

where L^b stands for the functional differential operator

$$\partial_\nu^y \frac{\delta}{\delta A_\nu^b(y)} + g f^{bdc} A_\nu^d(y) \frac{\delta}{\delta A_\nu^c(y)} \tag{9.106}$$

which it replaces. Hence, from the above covariant transformation,

$$L^b(y) J_\mu^a(x) = g f^{abc} J_\mu^c \delta^4(x-y). \tag{9.107}$$

The covariant divergence of the earlier equation can be calculated and yields

$$L^b(y)[\partial^\mu J_\mu^a + g f^{ade} A^{d\mu} J_\mu^e] = g f^{abc}[\partial^\mu J_\mu^c + g f^{cde} A^{d\mu} J_\mu^e]\delta^4(x-y). \tag{9.108}$$

If the covariant anomaly, i.e., the anomaly or covariant divergence of the covariantly regularized current is denoted by G_{cov}, this means

$$L^b(y) G_{cov}^a(x) = g f^{abc} \delta^4(x-y) G_{cov}^c(x). \tag{9.109}$$

This also leads to

$$L^a(x) G_{cov}^b(y) - L^b(y) G_{cov}^a(x) = -2g f^{abc} G_{cov}^c \delta^4(x-y). \tag{9.110}$$

Note that the operators $L^b(y)$ satisfy the commutation relation

$$[L^a(x), L^b(y)] = -g f^{abc} \delta^4(x-y) L^c. \tag{9.111}$$

The consistent anomaly is the anomaly or covariant divergence of a current obtained as the functional derivative of an effective action. Hence by applying the above equation on the effective action, we obtain

$$L^a(x) G_{con}^b(y) - L^b(y) G_{con}^a(x) = -g f^{abc} \delta^4(x-y) G_{con}^c(x), \tag{9.112}$$

where the consistent anomaly has been denoted by G_{con}. The equations satisfied by the two anomalies are different. This means that the covariant anomaly is not derivable from an effective action and the consistent anomaly is not gauge covariant. However, the difference is proportional to the anomaly, meaning that the nonintegrability or noncovariance is due to the anomaly itself.

An integrated form of the consistency condition is

$$\delta_u G_{con}^v - \delta_v G_{con}^u = ig G_{con}^{[u,v]}, \tag{9.113}$$

where the gauge variation with an infinitesimal function $u(x) = T^a u^a(x)$ is represented by

$$\delta_u \equiv \int d^4x \, u^a(x) L^a(x), \tag{9.114}$$

and the anomaly is smeared:

$$G_{con}^u \equiv \int d^4x \, u^a(x) G_{con}^a. \tag{9.115}$$

This equation can be seen to be satisfied by

$$G^u_{con} \propto \int d^4 x \, \text{tr}[u(x)\epsilon^{\mu\nu\rho\sigma}\partial_\mu(A_\nu\partial_\rho A_\sigma - \frac{ig}{2}A_\nu A_\rho A_\sigma)]. \qquad (9.116)$$

As it is not covariant, it cannot be written in terms of the field strength tensors. But it vanishes under the same condition as before, namely, the vanishing of d^{abc}. The normalization of the consistent anomaly is not fixed by the linear condition that it satisfies, but it can be calculated[6] from the covariant anomaly. The consistent anomaly is equal to

$$\int_0^g \frac{dg'}{g} G_{cov}(g') + \text{higher powers of } g. \qquad (9.117)$$

This yields the normalization

$$G^a_{con} = \frac{1}{24\pi^2}g^2\text{tr}[T^a\epsilon^{\mu\nu\rho\sigma}\partial_\mu(A_\nu\partial_\rho A_\sigma - \frac{ig}{2}A_\nu A_\rho A_\sigma)] \qquad (9.118)$$

because the covariant anomaly has a leading quadratic dependence on g, which produces a factor of $\frac{1}{3}$ and the denominator of 32 is converted to 8 when the F-s of the covariant anomaly are expressed in terms of A-s and the ϵ is used for simplification.

9.4.1   Quantizing chiral gauge theories with local gauge anomalies

We seek to study the quantization of the gauge field, the fermions in the chiral gauge theory being understood to have been quantized earlier, whereupon the anomaly in the theory arose.

As seen earlier, some classical symmetry currents cease to be conserved after quantization. In case the current is associated with a symmetry which is gauged, there may be a problem in the quantization of the gauge fields because the equations of motion of the gauge fields require the current to be covariantly conserved. Chiral gauge theories have anomalies unless the $d^{abc} = 0$ condition is satisfied. This can be resolved if the anomaly itself can be made to vanish by going to a submanifold of the classical phase space before quantization. Of course, there is a difference from theories with nonanomalous gauge currents. In those theories, there is gauge freedom, which means that the theories can be described in any of an infinite variety of gauges. This is not possible in a straightforward manner in anomalous gauge theories, where the gauge may have to be fixed by the anomaly.

If one follows the canonical procedure of quantization, there is no conceptual difficulty in quantizing these theories. The canonical method of quantization requires the determination of the momenta corresponding to the different

---

[6]H. Banerjee, R. Banerjee and P. Mitra, *Z. Physik* **C32**, 445 (1986).

field variables. As usual, A_0 has no canonical conjugate, so there is a primary constraint $\pi_0 = 0$. To preserve this constraint in time, it is necessary to have a further constraint, and this is how Gauss' law appears. In anomaly-free theories, no further constraint arises, and the above two constraints have vanishing Poisson brackets, i.e., are first class. In anomalous theories, the preservation in time of Gauss' law may require further constraints. Classical equations of motion suggest that the covariant time derivative of the Gauss law operator X^a vanishes. This is because of the covariant conservation of the current, which itself is required by the classical field equations for the gauge fields. However, in the presence of anomalies, this conservation is violated. The Gauss law constraint itself, which is a classical equation, is not expected to be an operator equation. In this situation, one finds

$$\partial^0 X^a + g f^{abc} A^{b0} X^c = g G^a. \tag{9.119}$$

This suggests a new constraint involving the anomaly. That in turn may give rise to a further constraint, and so on. If there are two new constraints, then these constraints are analogous to the gauge conditions that one has the freedom to choose in ordinary gauge theories, so one may say that the gauge is determined by the anomalous theory itself. Alternatively, the Poisson bracket of π_0 and the Gauss law operator can be nonvanishing. The closure of the set of constraints at the level of Gauss' law or after one additional set of constraints indicates fewer constraints than usual, resulting in the emergence of additional degrees of freedom. These may be thought of as would-be gauge degrees of freedom which have become physical because of the loss of gauge invariance. Whatever happens, the set of constraints has to be identified and imposed on the phase space, quantization being carried out thereafter in terms of the reduced degrees of freedom.

In the functional integral formulation of the theories, the full partition function of a gauge theory with fermions will be written as

$$Z = \int \mathcal{D}A Z[A], \tag{9.120}$$

where $Z[A]$ is obtained by functionally integrating the exponential of the classical action over the fermion fields.

If there were no chiral anomaly, $Z[A]$ would be gauge invariant. The presence of a gauge anomaly makes $Z[A]$ vary with gauge transformations of A:

$$Z[A^U] = e^{i\alpha(A,U^{-1})} Z[A], \tag{9.121}$$

where α, which is an integral representation[7] of the anomaly, has to satisfy the consistency requirements

$$
\begin{aligned}
\alpha(A, U_2^{-1} U_1^{-1}) &= \alpha(A^{U_1}, U_2^{-1}) + \alpha(A, U_1^{-1}), \\
\alpha(A, U^{-1}) &= -\alpha(A^U, U).
\end{aligned}
\tag{9.122}
$$

[7]L. D. Faddeev, *Phys. Letters* **145B**, 81 (1984).

In theories without gauge anomalies, the partition function factorizes into the volume of the gauge group and a gauge-fixed partition function, as we have seen before. Such a decoupling of gauge degrees of freedom may not occur if a local gauge anomaly is present. In this case, one has[8]

$$
\begin{aligned}
Z &= \int \mathcal{D}A Z[A] \\
&= \int \mathcal{D}A Z[A] \int \mathcal{D}U \delta(f(A^U)) \Delta_f(A) \\
&= \int \mathcal{D}U \int \mathcal{D}A Z[A^{U^{-1}}] \delta(f(A)) \Delta_f(A) \\
&= \int \mathcal{D}U \int \mathcal{D}A e^{i\alpha(A,U)} Z[A] \delta(f(A)) \Delta_f(A) \\
&\neq \left(\int \mathcal{D}U \right) Z_f.
\end{aligned}
\tag{9.123}
$$

In deriving the fourth equality, the noninvariance of $Z[A]$ under a gauge transformation has been used. This is the reason for the difference from the invariant situation. Sources have been suppressed for simplicity. Here U and A are coupled because of the anomaly term α. As indicated above, there are two possibilities: first, it may happen that there is a gauge function f_0 such that α vanishes in the special gauge $f_0 = 0$ and so the gauge degrees of freedom decouple in this gauge; alternatively, if there is no such gauge, the would-be gauge degrees of freedom become physical. In this latter case one can fix a gauge in an enlarged theory where the field U in the above expressions is also taken as a dynamical field. By using the consistency condition (9.122) for α, the product $Z[A, U] \equiv e^{i\alpha(A,U)} Z[A]$ can be seen to be invariant under gauge transformations of A if U is appropriately transformed at the same time:

$$
Z[A^V, UV] = Z[A, U].
\tag{9.124}
$$

This is the procedure of making an anomalous theory gauge invariant by introducing what is called a Wess–Zumino field to cancel the gauge variation. The new theory with

$$
Z = \int \int \mathcal{D}U \mathcal{D}A Z[A, U] \delta(f(A)) \Delta_f(A)
\tag{9.125}
$$

can be sought to be treated by standard methods. However, the knowledge of nonabelian vector gauge theories cannot be simply extended to these complicated theories. Experience gained from two dimensions[9] indicates that there may be some theories, or more precisely some regularizations, for which unitarity is violated. Other regularizations may work. Note that the regularization enters the picture through the form of the anomaly.

---

[8]K. Harada and I. Tsutsui, *Phys. Letters* **183B**, 311 (1986).
[9]R. Rajaraman, *Phys. Letters* **154B**, 305 (1985).

9.4.2 Global anomaly

There are anomalies of two different types to be considered in the quantization of chiral gauge theories. What we discussed above is the case of common or local anomalies. The second kind of anomaly — the so-called global kind — is more subtle. Here the gauge current is conserved, but the group of time independent gauge transformations is not simply connected. This has consequences for quantization in the so-called temporal gauge. One obtains a representation of the Lie algebra of the group of time independent gauge transformations in the Hilbert space of states, but this provides in general only projective or multiple valued representations of the group itself. When the fermion content is such that the representation is not a true one, there is no state in the Hilbert space which is invariant under the group, so that the subspace of states obeying Gauss' law is trivial. This problem can, however, be avoided by fully fixing the gauge. The difference between theories with global anomalies and anomaly-free theories turns out to be slight.

The impossibility of imposing Gauss' law is specific to the temporal gauge and is to be contrasted with canonical quantization, where constraints and gauge conditions are imposed at the classical level and quantization is carried out on the nonsingular theory. Gauss' law and the gauge condition reduce the phase space. The dynamical system that remains can be quantized as usual. Gauss' law becomes an operator equation in the Hilbert space, which does *not* carry any nontrivial representation of the gauge group or its Lie algebra.

If there is a global anomaly, there is a special form of α. One way of characterizing a theory with a global anomaly is to say that the full group of time dependent gauge transformations is disconnected. Thus there is a possibility of distinguishing between transformations not connected to the identity and ones obtainable from the identity by a sequence of infinitesimal transformations. It is only under the former, i.e., the large gauge transformations, that $Z[A]$ does not stay invariant in these theories. To be precise, the transformation is given by

$$Z[A^U] = e^{i\gamma(U)} Z[A], \tag{9.126}$$

where $\gamma(U)$ vanishes only for gauge transformations U connected to the identity.

The partition function in this case seems to factorize:

$$Z = \int \mathcal{D}U e^{-i\gamma(U)} \int \mathcal{D}A Z[A]\delta(f(A))\Delta_f(A). \tag{9.127}$$

But one must be careful. The phase factors form a representation of the group, so

$$\int \mathcal{D}U e^{-i\gamma(U)} = \int \mathcal{D}(UV)e^{-i\gamma(UV)} = e^{-i\gamma(V)} \int \mathcal{D}U e^{-i\gamma(U)}, \tag{9.128}$$

where a fixed element V of the gauge group has been used. If it is not connected to the identity, the left and right sides seem to differ by a phase factor, indicating that $\int \mathcal{D}U e^{-i\gamma(U)}$ must vanish. This implies that the partition function Z vanishes. In fact, this was an argument against the definability of such theories. However, one is really interested in the expectation values of gauge invariant operators:

$$\frac{\int \mathcal{D}AZ[A]\mathcal{O}}{\int \mathcal{D}AZ[A]} = \frac{\int \mathcal{D}U e^{-i\gamma(U)} \int \mathcal{D}AZ[A]\delta(f(A))\Delta_f(A)\mathcal{O}}{\int \mathcal{D}U e^{-i\gamma(U)} \int \mathcal{D}AZ[A]\delta(f(A))\Delta_f(A)}. \qquad (9.129)$$

The right side is of the form $0/0$ because of the presence of the factor $\int \mathcal{D}U e^{-i\gamma(U)}$ in the numerator and the denominator. One can interpret this ratio in a sensible way by removing[10] this common vanishing factor.

$$< \mathcal{O} >= \frac{\int \mathcal{D}AZ[A]\delta(f(A))\Delta_f(A)\mathcal{O}}{\int \mathcal{D}AZ[A]\delta(f(A))\Delta_f(A)}. \qquad (9.130)$$

This expression is precisely what one gets in the canonical approach to quantization. We have been considering the Lagrangian functional integral, where the singular nature of the Lagrangian is ignored and all degrees of freedom, physical or unphysical, integrated over. In the canonical approach, on the other hand, the gauge degrees of freedom are removed by fixing the gauge at the classical level and only the physical part of the theory is quantized. The functional integration is then over only the physical fields. There are both ordinary fields and conjugate momenta to be integrated over, but the latter are easily integrated over, resulting in functional integrals leading to the gauge fixed expression given in (9.130). This is reached without having to go through the full partition function which is used in the Lagrangian approach and appears to lead to a problem by happening to vanish for theories possessing global anomalies.

This resolution of the problem does not mean that there is no trace whatsoever of the global anomaly. An interesting consequence of the disconnectedness of the gauge group is that gauge fixing functions f can be classified. Two functions f and f' may be said to belong to the same class if one can find a gauge transformation connected to the identity to go from a gauge field configuration satisfying one gauge condition to one satisfying the other. In this situation, Z_f and $Z_{f'}$ can be shown to be equal by simple manipulations. In general, however, the transformation that is needed to pass from a gauge field configuration consistent with f to one consistent with f' is not connected to the identity. To see what happens in this situation, one has to go through manipulations similar to those used in anomaly-free theories to show that the

---

[10]P. Mitra, *Letters Math. Phys.* **31**, 111 (1994).

gauge fixed partition function is the same for different gauge functions. Thus,

$$
\begin{aligned}
Z_f &= \int \mathcal{D}A Z[A]\delta(f(A))\Delta_f(A) \\
&= \int \mathcal{D}A Z[A]\delta(f(A))\Delta_f(A) \int \mathcal{D}U \delta(f'(A^U))\Delta_{f'}(A) \\
&= \int \mathcal{D}U \int \mathcal{D}A Z[A]\delta(f(A))\Delta_f(A)\delta(f'(A^U))\Delta_{f'}(A) \\
&= \int \mathcal{D}U \int \mathcal{D}A Z[A^{U^{-1}}]\delta(f(A^{U^{-1}}))\Delta_f(A)\delta(f'(A))\Delta_{f'}(A) \\
&= \int \mathcal{D}A Z[A]\delta(f'(A))\Delta_{f'}(A) \int \mathcal{D}U e^{-i\gamma(U)}\delta(f(A^{U^{-1}}))\Delta_f(A).
\end{aligned}
$$

$$(9.131)$$

Were it not for the phase factor $e^{-i\gamma(U)}$, the last integral would be the identity and the right side would reduce to the gauge-fixed partition function for the gauge function f'. The two integrals appear to be coupled here. But that is not really the case. Although different gauge field configurations have to be integrated over, only those are relevant for which both $f'(A)$ and $f(A^{U^{-1}})$ vanish, and the second condition picks out one U for each A satisfying the first condition. As A changes continuously — the spacetime manifold is taken to be compactified — U varies in a fixed homotopy class, so that $\gamma(U)$, which depends only on the class, remains unchanged. Consequently, the factor can be pulled out and one can write

$$
Z_f = e^{-i\gamma(U_0)} Z_{f'}, \tag{9.132}
$$

where U_0 is an element of the relevant homotopy class, which is determined by the gauge functions f and f'. It is through these factors that theories with global anomalies differ from anomaly-free theories. But these factors occur only in the partition functions and clearly cancel out in the expectation values of gauge invariant operators, so that Green functions of gauge invariant operators are fully gauge independent.

Are theories with global anomalies unitary and renormalizable? They indeed are, as may be understood by recalling that the gauge current is conserved in these theories — if it were not, one would be dealing with a theory with the local kind of anomaly instead of the global kind. The invariance under infinitesimal gauge transformations which exists in these theories implies a BRS invariance within each gauge fixed version. This, together with the global gauge invariance of Green functions of formally gauge invariant objects, may be used to study unitarity and renormalizability. Indeed, all of standard perturbative gauge theory relies only on invariance under infinitesimal gauge transformations, i.e., under the group of gauge transformations connected to the identity. The theories under discussion do possess this invariance: $\gamma(U)$ is zero for all U connected to the identity.

There is a mathematical issue which may be raised: the possibility of choosing a gauge condition in these theories. There is, in fact, a problem, but it applies to all gauge theories irrespective of whether they are afflicted by anomalies of any kind. There is a theorem which says that gauges cannot be chosen in a smooth way. However, for the construction of functional integrals, it is sufficient to have piecewise smooth gauges. This is the situation even for theories without anomalies.

Index